Linear and Nonlinear System Modeling

Scrivener Publishing
100 Cummings Center, Suite 541J
Beverly, MA 01915-6106

Publishers at Scrivener
Martin Scrivener (martin@scrivenerpublishing.com)
Phillip Carmical (pcarmical@scrivenerpublishing.com)

Linear and Nonlinear System Modeling

Edited by

Tamal Roy

Suman Lata Tripathi

and

Souvik Ganguli

Scrivener Publishing

WILEY

This edition first published 2024 by John Wiley & Sons, Inc., 111 River Street, Hoboken, NJ 07030, USA and Scrivener Publishing LLC, 100 Cummings Center, Suite 541J, Beverly, MA 01915, USA

For more information about Scrivener publications please visit www.scrivenerpublishing.com.

Wiley Global Headquarters
111 River Street, Hoboken, NJ 07030, USA

For details of our global editorial offices, customer services, and more information about Wiley products visit us at www.wiley.com.

Library of Congress Cataloging-in-Publication Data

ISBN 978-111984-742-7

Front cover images supplied by Adobe Firefly
Cover design by Russell Richardson

Set in size of 11pt and Minion Pro by Manila Typesetting Company, Makati, Philippines

Printed in the USA

10 9 8 7 6 5 4 3 2 1

Contents

Preface

Mathematical modeling of the system is extracting the essence of practical problems into systematic mathematical language. In system modeling, mathematical expression deals with modeling and its applications. It is characterized by how a modeling competency can be categorized and how its activity can contribute to building up these competencies. Mathematical modeling of a practical system is an attractive field of research and a grown-up subject with a variety of applications. The main objective of mathematical modeling is to predict the behavior of the system under different operating conditions, to design and implement efficient control strategies to achieve the desired performance. A considerable research effort has been directed to the development of a model, which must be understandable and easy to analyze. It is a very difficult task to develop mathematical modeling of complicated practical systems considering all its possible high-level non-linearity and cross-couple dynamics. Although mathematical modeling of nonlinear systems sounds quite interesting, it is difficult to formulate the general solution to analyze and synthesize nonlinear dynamical systems.

This book covers the different features of the mathematical modeling of real physical systems like grid-connected hybrid power system, microgrid network, harmonics modeling of the power system, active power filter, photovoltaic panels, current transformer, induction motor drive, DC/DC converter, and lithium-ion batteries. Most of the real practical systems are nonlinear in nature and synergetic integration of the various sub systems connected in such a manner to perform the desired task. Therefore, any attempt to derive the mathematical modeling of such real physical systems is a challenging task. This book covers the derivation and implementation of the mathematical modeling of such real physical systems from different fields of engineering.

We hope that the contents of the book would be useful for teaching and research. Any suggestions for further improvements to the book are most welcome. We are grateful to all who encouraged and helped us directly or indirectly during the preparation of the manuscript.

Tamal Roy
Suman Lata Tripathi
Souvik Ganguli
Kolkata, India
June 2024

1

Assessment of Faults in Hybrid System Connected with Main Grid

Aveek Chattopadhyaya[1]*, Niladri Mukherjee[2] and Surajit Chattopadhyay[3]

[1]Department of Electrical Engineering, GNIT, Panihati, Kolkata, India
[2]Department of Electrical Engineering, S.K.F.G.I., Mankundu, Hooghly, India
[3]Department of Electrical Engineering, GaniKhan. C. Institute of Engineering and Technology, Malda, India

Abstract

In this chapter, fault assessment in grid-connected microgrid (solar and wind-based hybrid system) was performed by signal processing-based waveform analysis. For this analysis, wind and solar system-based grid-connected microgrid (hybrid system) is considered. System current waveforms were captured for fault identification. Total harmonic distortion (THD) based on fast Fourier transform (FFT) and inter-harmonic group analysis were considered as tools to identify the presence of different harmonics in non-identical environments. In Park plane, one space vector was also calculated from the captured three-phase current signals. Different parameters based on discrete wavelet transform (DWT) were also computed from space vector for assessing the symmetrical and unsymmetrical faults in hybrid system, which is connected to the main grid.

Keywords: THD, FFT, DWT, skewness, kurtosis, mean RMS

1.1 Introduction

The demand for renewable energy resources has escalated significantly over the past few years, due to the fact that the fossil fuels will run out in the near future and the harmful environmental effects of the fossil fuels. In hybrid system, wind & solar systems are used for generating power where

**Corresponding author*: chatterjee58@gmail.com

Tamal Roy, Suman Lata Tripathi and Souvik Ganguli (eds.) Linear and Nonlinear System Modeling, (1–14) © 2024 Scrivener Publishing LLC

wind turbines generators are used for wind generation and the solar panels are used for solar generation. The power generation of solar and wind system strongly depends on the weather conditions. Distributed generation plays a vital role in today's huge energy demand scenario. Hybrid generation system is a system, which is having renewable/non-renewable sources. Hybrid systems are interconnected with renewable/non-renewable sources to supply power to local load and grid/micro grid. Different issues like control, optimal sizing, grid connection of hybrid systems having solar photovoltaic (PV), wind power, fuel cells etc. have been vividly ventilated by the researchers [1]. Different issues regarding benefits and drivers of 'hybrid renewable energy systems (HRES)' for off grid power generation has been discussed in [2]. A simulink model is developed of grid coupled hybrid system in MATLAB where the hybrid system consists of wind and photovoltaic systems [3]. To simulate a hybrid system some basic systems are required. In PV/wind, hybrid systems three important parts are needed [4]. Firstly, PV/wind system, wind energy conversion (WEC) system, power electronics devices and MPPT algorithm [3]. The hybrid system encompasses PV-wind system, DC/DC converter and energy (wind) conversion system [4]. E. Natsheh *et al.* proposed a control strategy for the optimization of smart grid performance where a hybrid system is connected with a grid [5]. In [6], a modular PV and fuel cell based grid connected hybrid system has been developed in MATLAB and '160 Wp solar module' and '5 kW solid oxide fuel cell (SOFC)' was used for simulation. N. T. Pathan *et al.* discussed steady state performance of a hybrid system connected with main grid [7]. In this study permanent - magnet - synchronous – generator has been used for wind turbine. Two different system (wind and solar) are synchronized to supply a common grid and a local load. The output of the system fed to a common grid and a local load while solar system generates dc power and wind system produces pulsating AC. The main difficulty is synchronization. The output power from wind system is converting into DC then connected with solar system. Afterword it is fetched to a PWM inverter and finally synchronized [9]. For the reliable operation of grid connected microgrid system, it is very much obvious that, different faults should be detected properly as early as possible. In [8], authors designed and simulated and studied hybrid system consists of wind and solar in isolated mode [8]. A. González *et al.* suggested sensitivity analysis based optimally sizes of 'grid-connected photovoltaic–wind power systems' in [10]. Using this system, cost of electricity has been lowered as compared to purchase of electricity. S. Swain proposed an Active Crowbar Protection (ACB_P) system to improve the system's power quality by enhancing the DFIG's fault-ride through (FRT) capacity. The suggested scheme's functionality is confirmed by running simulations on a 1.7

MVA "wind energy systems" based on "Doubly Fed Induction Generators (DFIG)" under various short circuit fault scenarios in MATLAB [11]. In this chapter, consequences and identification of faults are discussed thoroughly based on harmonics analysis. The different types of faults in grid integrated hybrid system have been investigated by the inter-harmonics; total harmonic distortion (THD), Park plane and DWT based skewness, kurtosis, root-mean-square (RMS) and mean value analysis. Here, in grid side non-identical faults have been contemplated for analysis. Three-phase system current has been captured at 'point of common coupling (PCC)' and normalized and then it has been processed to assess the different conditions.

1.2 Hybrid System Connected with Main Grid

The single line diagram of grid-connected hybrid system is delineated in the above figure (Figure 1.1). The hybrid system is comprised of a wind generation and solar generation. A 100-KW photovoltaic module (*i.e.*, PV array) consisting of 96 cells per module with 66 parallel string, having 5 series-connected modules per string are connected, and a 9-MW wind farm, having six numbers of 1.5-MW wind turbines were used for making the hybrid system. The solar system yields 64.2 V, which was boost up near to 500 V by boost converter. Here, the wind generator output nominal voltage is 575 V. An induction generator (wound rotor) was used in doubly-fed-induction-generator (DFIG) in wind generation. Wind generator output voltage is pulsating ac. The output voltage of wind system was fed to a universal bridge rectifier to produce DC output. Rectifier output is

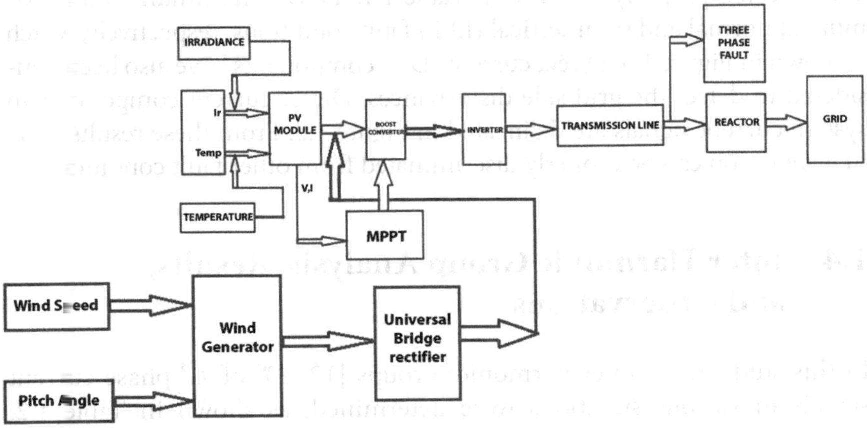

Figure 1.1 Single line representation of hybrid system connected with main grid.

then directly connected to the output of the boost converter of the solar system for proper synchronization because both the output is DC. In DC, the synchronizing process is easier than the ac synchronizing process because in the ac system, the main concerns are phase sequence and frequency mismatch. To overcome these problems, synchronization was done through DC. The combined output of the hybrid system was then fed to the inverter. Inverter output voltage was then stepped up to 25 kV through a step-up transformer. The output of the transformer was transmitted through 15-km line and fed to a 25-kV grid. Here, non-identical faults were contemplated in the grid side. In fault and normal situations, all three-phase currents of bus (after transformer) was considered at proper sampling frequency for analysis. The three-phase current was captured and normalized then using Park transformation matrix, one space vector was found. After finding out the current space vector (in Park plane), it was processed by DWT. In DWT, the mother wavelet was considered as 'db4' as it is well suited for this analysis. From the DWT coefficients, different parameters like kurtosis, mean, RMS, and skewness were calculated in different conditions to detect and identify the grid side disturbances. Fast Fourier transform (FFT) and inter-harmonics groups were also computed on captured system current signals to assess the abnormal conditions.

1.3 FFT Results in Different Conditions, Respective Bar Diagram, and Observations

From THD analysis based on FFT, different conditions can be discriminated and detected properly [12, 13]. In Table 1.1, THD is minimum and maximum at normal and symmetrical (LLL) fault conditions, respectively, which is shown in Figure 1.3. Direct current (DC) components have also been considered to detect the grid side disturbances. Direct current components in system current signals are delineated in Figure 1.2. From these results, normal condition can be properly discriminated from other fault conditions.

1.4 Inter-Harmonic Group Analysis, Results, and Observations

In this study, three inter-harmonics groups [12, 13] of 'C' phase current signals in various situations were determined, as shown in Table 1.2. Maximum group values of *inter-harmonics* were found in *LLL* and *LL* fault,

Table 1.1 THD of current for C phase.

Condition	DC	1st	2nd	3rd	4th	5th	6th	7th	8th	9th	10th	11th	THD
Normal	0.23	1	0.02	0.06	0.01	0.05	0.08	0.36	0.08	0.08	0.02	0.01	1.04
L-G	23.67	1	1.50	1.01	0.75	0.61	0.51	0.45	0.35	0.32	0.29	0.27	9.26
L-L-G	13.90	1	0.90	0.61	0.47	0.41	0.40	0.62	0.37	0.20	0.17	0.15	6.02
L-L-L	136.01	1	0.01	5.42	7.80	11.82	21.07	58.56	44.97	21.11	13.74	10.24	287.38
L-L	38.49	1	2.62	3.74	3.98	5.40	8.51	7.61	6.84	6.28	3.82	2.63	99.20

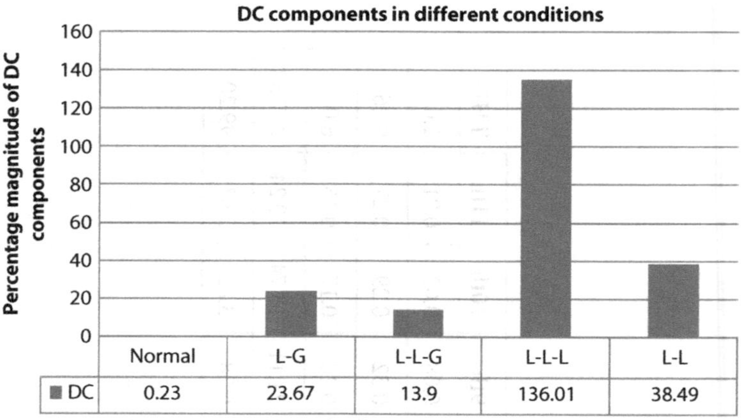

Figure 1.2 DC components in different conditions.

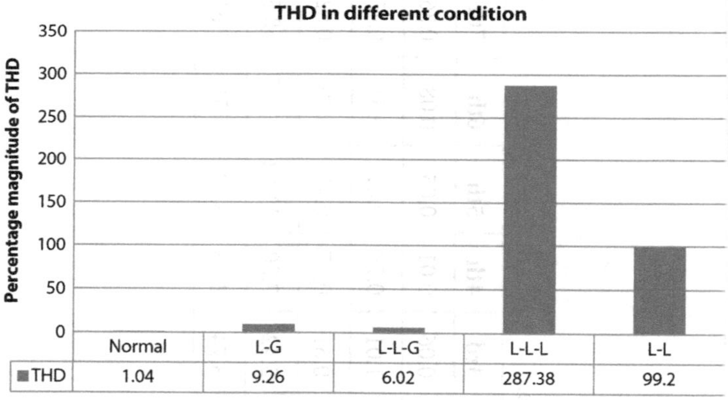

Figure 1.3 THD in different conditions.

Table 1.2 Inter-harmonics in current signal (C phase).

Inter-harmonic group	Normal or no-fault condition	LG condition	LLG condition	LL condition	LLL condition
IG1	12.00	13.32	12.48	14.72	20.86
IG2	12.00	12.47	12.17	15.04	19.33
IG3	12.00	12.24	12.09	12.01	21.57

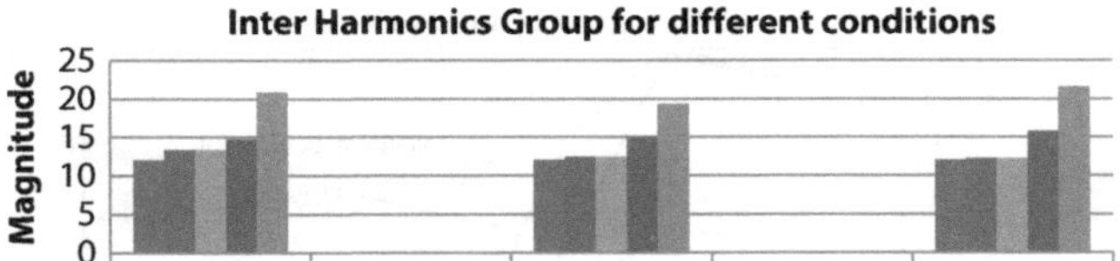

	Ig1	Ig2	Ig3
Normal	12	12	12
LG	13.32	12.47	12.243
LLG	13.32	12.4715	12.25
LL	14.724	15.0045	15.8348
LLL	20.86	19.3341	21.5753

Figure 1.4 Group analysis (inter-harmonics) of 'C' phases current in non-identical situations.

whereas minimal values were found in normal environment. The outcomes are illustrated in Figure 1.4. Different abnormal circumstances in the grid side were analyzed from there.

1.5 Statistical Parameter Analysis Based on Discrete Wavelet Transform, Results, and Observations

The currents in all three phases of the system were recorded and then normalized. One space vector was found out using the Park transformation matrix. After determining the current space vector (in Park plane), the 'db4' mother wavelet-based DWT is used to deconstruct it. Different metrics such as kurtosis, mean, RMS, and skewness were derived from DWT in different circumstances, as shown in the accompanying figures [12–16]. Kurtosis and mean values of DWT coefficients (approximation) are depicted in Figure 1.5(a) and (b). The highest difference between normal and L-G fault conditions is shown in Figure 1.5(a), whereas the minimum difference is shown in Figure 1.5(b). Figure 1.6(a) and (b) shows the findings of RMS and skewness values of DWT coefficients (approximation) in different situations, respectively. In different DWT levels, zig-zag discrepancies between normal and other fault circumstances are identified in Figure 1.6(a). Up to DWT fragmentation level three (3), the highest difference between *normal* and *L-L-L fault* environments may be seen in Figure 1.6(b). In nature, the disparity has been shrinking since level three. From level one to nine, the smallest difference was noticed between other faults (*L-L, L-L-G, L-G fault situations*) and conventional condition. The findings of kurtosis and mean of detail coefficients at non-identical situations are defined in Figure 1.7(a) and (b). Figure 1.7(a) and (b) shows a zig-zag

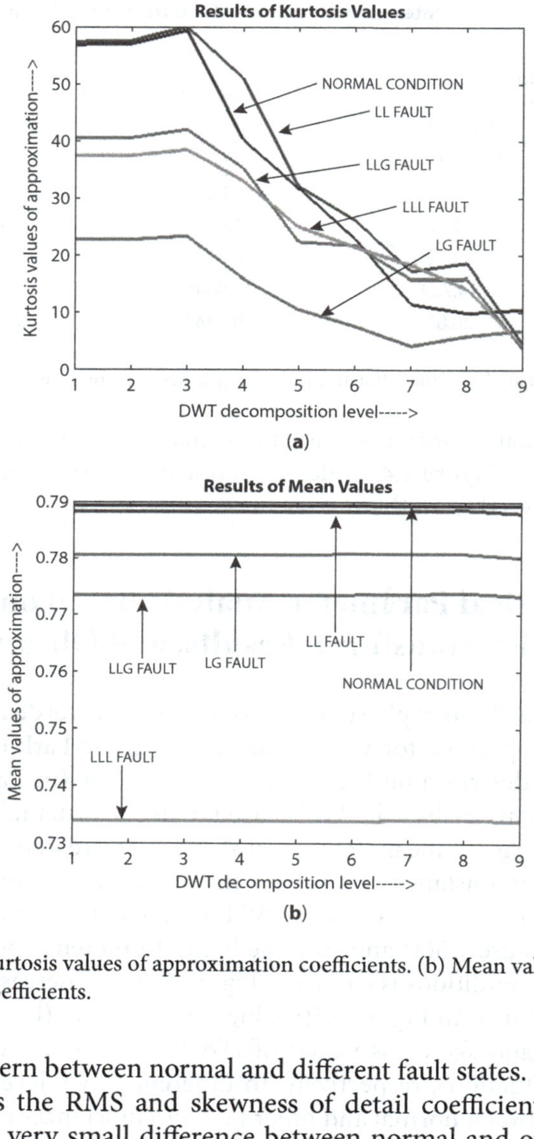

Figure 1.5 (a) Kurtosis values of approximation coefficients. (b) Mean values of approximation coefficients.

difference pattern between normal and different fault states. Figures 1.8(a) and (b) shows the RMS and skewness of detail coefficients results. Up to level two, a very small difference between normal and other fault circumstances are shown in Figure 1.8(a). The difference between level two and four has remained constant for L-G problems, but it has increased in character for other types of conditions. The highest and minimum differences between conventional and *L-G* and *L-L-L* fault situations have been reported. Figure 1.8(b) shows some zig-zag changes of skewness values in different conditions.

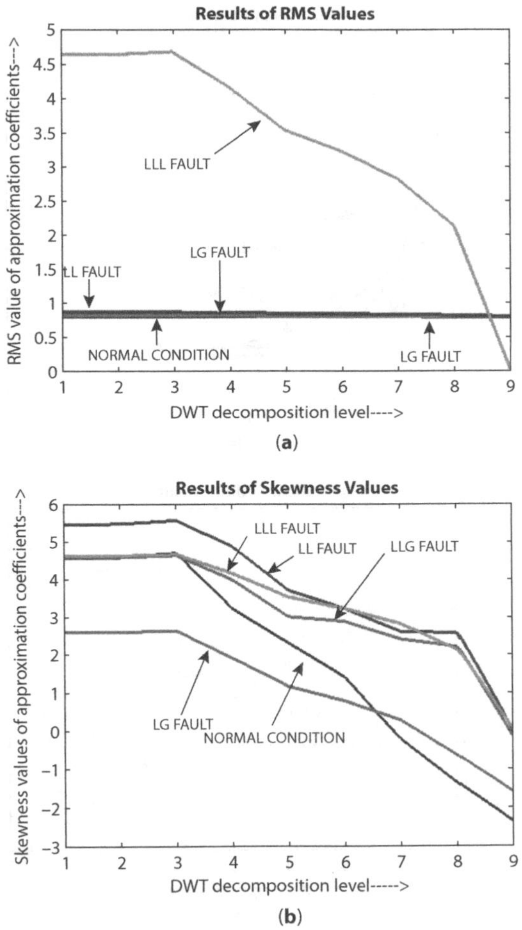

Figure 1.6 (a) RMS values of approximation coefficients. (b) Skewness values of approximation coefficients.

1.6 Algorithm to Determine Non-Identical Conditions

The following is an algorithm for assessing different situations in a hybrid system (*grid-connected*) and could be implemented for protection of grid-connected microgrid:

- First, the three phases of the current signal will be sampled at the appropriate sampling frequency.

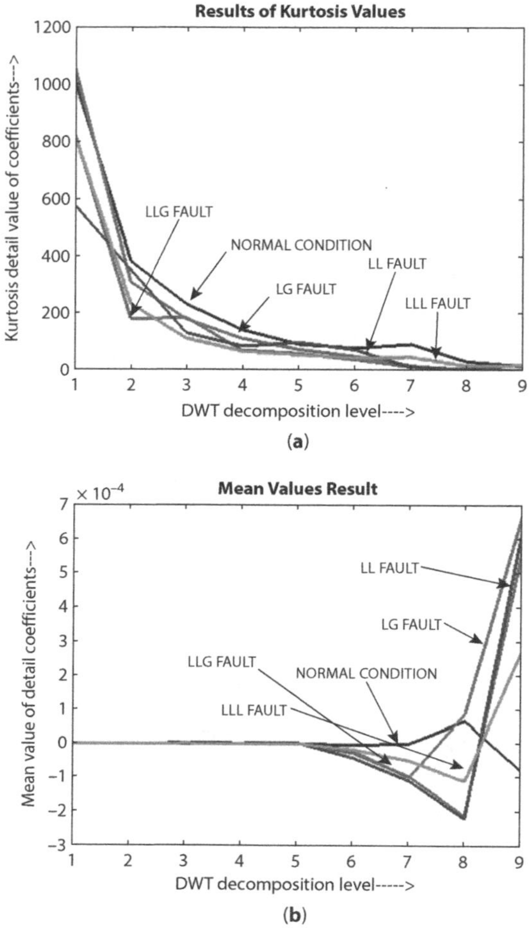

Figure 1.7 (a) Detail coefficients kurtosis values. (b) Mean values of detail coefficients.

- The three-phase current signal must then be transformed into the Park plane using the Park transformation matrix.
- One current space vector will be calculated in the Park plane.
- The current space vector will then be decomposed using a DWT based on a 'dB4' mother wavelet.
- Then, at various DWT decomposition levels, various parameters such as skewness, kurtosis, and so on, will be calculated.
- Different fault conditions will be assessed by comparing all of the 'above-calculated' parameters to the normal condition.

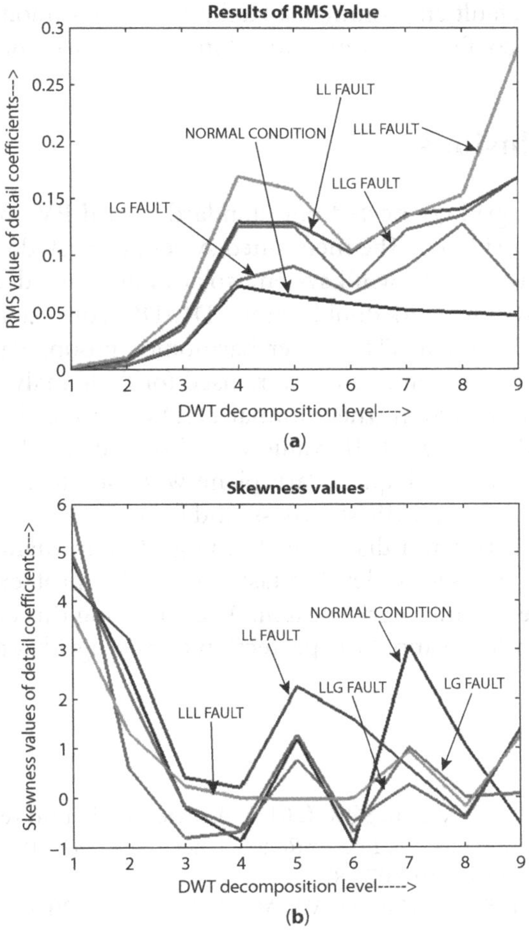

Figure 1.8 (a) RMS from DWT coefficients (detail). (b) Skewness values of detail coefficients.

1.7 Specific Outcome of This Chapter

The Park plane, THD based on FFT, inter-harmonics group, and parameters estimations based on DWT were used to investigate the non-identical conditions in hybrid electrical system, which is connected in main grid. The normal condition had the lowest THD value and the LLL fault situation had the highest THD. In another analysis, the minimum inter-harmonics were noticed in the conventional situation and the maximum in the *LLL* and *LL* fault environments. Highest and continuous deviations in approximation coefficients were noticed in DWT-based analysis.

1.8 Conclusions

In this chapter, grid connected one standard hybrid PV and wind power system were considered. The mentioned system consisted of four important segments, which are: wind system, solar system, wind energy conversion (WEC) system, and double-input DC/DC converter. Fast Fourier transform (FFT)-based THD, inter-harmonics group, Park plane, and DWT-based different parameters were used for fault analysis in wind and solar-based hybrid system. The minimal THD was noticed at conventional situation and the highest THD value was observed at L-L-L symmetrical fault situation. In this chapter, Park plane was used to extract the space vector and then mean, RMS, skewness, and kurtosis were computed based on DWT to identify and discriminate the grid abnormalities. Discussed techniques can be used to develop fast and reliable protection scheme in any grid-connected microgrid system. After studying this chapter, readers can gather knowledge about the protection of microgrid system.

References

1. Krishna, S, K., & Kumar, S., K. (2015). A review on hybrid renewable energy systems. *Renew. Sustain. Energy Rev.*, Science Direct, 52, 907–916.https://doi.org/10.1016/j.rser.2015.07.187.
2. Mohammed, S., Y., Mustafa, W., M., & Bashir, N. (2014). Hybrid renewable energy systems for off-grid electric power: Review of substantial issues. *Renew. Sustain. Energy Rev.*, Science Direct, 35, 527–539. https://doi.org/10.1016/j.rser.2014.04.022.
3. Rana, A., & Ilyas, Md. (2015). Implementation of a wind/pv hybrid system using matlab/simulink. *Int. J. Adv. Res. Electr. Electron. Instrum. Eng.*, 4(7), 5941-5948. https://doi.org/10.15662/ijareeie.2015.0407019.
4. Harini, M., Ramaprabha, R., & Mathur, L., B. (2012). Modeling of grid connected hybrid wind/pv generation system using matlab. *ARPN J. Eng. Appl. Sci.*, 7(9), 1157-1161.ISSN 1819-6608.
5. Natsheh, M., E., Albarbar, A., & Yazdani, J. (2011). Modeling and control for smart grid integration of solar/wind energy conversion system. *2nd IEEE PES International Conference and Exhibition on Innovative Smart Grid Technologies (ISGT Europe)*, Manchester, UK, 1-8. https://doi./org/10.1109/ISGTEurope.2011.6162643.

6. Bayrak, G., & Cebeci, M. (2013). Grid connected fuel cell and PV hybrid power generating system design with Matlab Simulink. *Int. J. Hydrog.*, Science Direct, 39(16), 8803-8812.https://doi.org/10.1016/j.ijhydene.2013.12.029.
7. Pathan, T., N., & Adhau, P. S., & Sable, M. (2017). MPPT for grid connected hybrid wind driven PMSG - solar PV power generation system with single stage converter. *J. Electr. Power Syst. Eng.*, 3(1), 41-49.
8. Jyothi, A., S., P., Lakshmanrao, Kurian, P., C., & Singh, K., B. (2014). Design and simulation of wind and solar hybrid system under isolated mode of operation. *Int. J. Ind. Electron. Electr. Eng.*, 2(3), 23-27.
9. Ren, H., Wu, Q., Gao, W., & Zhou, W. (2016). Optimal operation of a grid-connected hybrid PV/fuel cell/battery energy system for residential applications. *Energy*, Science Direct. 113, 702-712. https://doi.org/10.1016/j.energy.2016.07.091.
10. González, A., Riba, R., J., Rius, A., & Puig, R. (2015). Optimal sizing of a hybrid grid-connected photovoltaic and wind power system. *Appl. Energy*, Science Direct. 154, 752–762. https://doi.org/10.1016/j.apenergy.2015.04.105.
11. Swain, S., & Ray, K., P. (2017). Short circuit fault analysis in a grid connected DFIG based wind energy system with active crowbar protection circuit for ride through capability and power quality improvement. *Electrical Power and Energy Systems*. 84, 64 –75. https://doi.org/10.1016/J.IJEPES.2016.05.006.
12. Mukherjee, N., Chattopadhyaya, A., Chattopadhyay, S., &Sengupta, S. (2020). Discrete Wavelet Transform and Stockwell Transform based Statistical Parameters Estimation for Fault Analysis in Grid Connected Wind Power System. *IEEE Syst. J.* 14(3), 4320 – 4328. https://doi.org/10.1109/JSYST.2020.2984132.
13. Dutta, S., Chattopadhyaya, A., Chattopadhyay, S., & Das, A. (2021). Harmonic Distortion, inter-harmonic Group Magnitude and Discrete Wavelet Transform based Statistical Parameter Estimation for Line to Ground Fault Analysis in Microgrid System. *MFIIS-2020*, Published in IET Digital Library. 177 – 184. https://doi.org/10.1049/icp.2021.1087.
14. Datta, S., Chattopadhyaya, A., Chattopadhyay, S., & Das, A. (2020). Line to ground and line to line fault analysis in IEEE standard 9 bus system. *AMSE-IIETA Journal* 19(504(2A)), 10-18. https://doi.org/10.18280/mmc_a.931-402.
15. Chattopadhyay, S., Chattopadhyaya, A., & Sengupta, S. (2014). Measurement of harmonic distortion and skewness of stator current of induction motor at crawling in clarke plane. *IET Sci. Meas. Technol.* 8(6), 528-536. https://doi.org/10.1049/iet-smt.2013.0082.
16. Chattopadhyay, S., Chattopadhyaya, A., & Sengupta, S. (2014). Analysis of stator current of induction motor used in transport system at single phasing by measuring phase angle, symmetrical components, skewness, kurtosis and harmonic distortion in Park plane. *IET Electr. Syst. Transp.*, 4(1), 1 – 8. https://doi.org/10.1049/iet-est.2012.0048.

2

Diversified Harmonics Modeling for Power System Stability Analysis

Tamal Roy[1], Debopoma Kar Ray[1] and Surajit Chattopadhyay[2]*

[1]Dept. of Electrical Engineering, MCKVIE, West Bengal, India
[2]Dept. of Electrical Engineering, GKCIET, West Bengal, India

Abstract
This chapter presents diversified harmonics modeling for power system networks. The chapter starts with a description of a simplified steady-state stability model. Network stability is classified based on load change and disturbance. Then the presence of harmonics is introduced step by step. Then a brief discussion is presented on how the model will look in the presence of harmonics. The change of operating points due to harmonics is discussed in a case study.

Keywords: Harmonics, harmonic operating point (HOP), harmonic power, load angle, maximum power, power equation, power system, stability

2.1 Introduction

The term 'stability' in the power system indicates the state of 'synchronism' maintained during operation. Power system networks involve synchronous generators along with other types of generation units. Moreover, they feed power to a variety of loads of diverse and dynamic characteristics. Also, power system networks face different types of faults and other types of disturbances. All these make the operation in synchronism a very difficult and challenging task to be maintained. The maximum power that can be delivered to maintain stability is known as the stability limit.

**Corresponding author*: surajitchattopadhyay@gmail.com

Tamal Roy, Suman Lata Tripathi and Souvik Ganguli (eds.) Linear and Nonlinear System Modeling, (15–32) © 2024 Scrivener Publishing LLC

2.2 Classification

System characteristics are greatly influenced by load condition, disturbance, method, and capacity to maintain stable operation. Concerning these, power system stability (PSS) is classified into three categories, as follows:

a. Steady-state stability
b. Transient stability
c. Dynamic stability

2.2.1 Steady-State Stability

Power system stability (PSS) under small disturbances followed by normal load conditions is known as steady-state stability. The systems are considered under automatic voltage control (AVC). Maximum power transfer is an important criterion for the study of steady-state stability. Maximum power point in power versus load angle characteristics is treated as the stability limit of steady-state stability. Beyond this limit, the system fails to operate in synchronism and fails to maintain stability.

2.2.2 Transient Stability

Power system stability (PSS) under large disturbance to bring the operation of the system in 'synchronism' is known as transient stability. A sudden large disturbance can cause a loss of stability if it fails to bring back the synchronism by reducing oscillation. Loss of stability may occur below the steady-state stability limit. Transient stability deals with the properties of the generator for reducing oscillation before the governor starts acting. Various methods which improve power system transient stability are [1–9]:

i. Improved steady-state stability
   a) Higher system voltage levels
   b) Additional transmission line
   c) Smaller transmission line series reactance
   d) Smaller transfer leakage reactance
   e) Series capacitive transmission line compensation
   f) Static var compensators and flexible AC transmission systems (FACTs)
ii. High speed fault clearing
iii. High speed reclosure of circuit breaker

iv. Single pole switching
v. Large machine inertia, lower transient reactance
vi. Fast responding, high-gain exciter
vii. Fast valving
viii. Breaking resistor

2.2.3 Dynamic Stability

Dynamic PSS refers to bringing the system's operation in 'synchronism' at the new or same equilibrium point of oscillation by governor action. It comes after the transient stability period or rather when the system fails to restore stability in its transient stability period. Dynamic stability depends on various factors like faults, breaking operations, and fault clearance.

Power system stability is mainly concerned with rotor stability analysis. For this, various assumptions are needed, such as:

- For stability analysis, balanced three-phase system and balanced disturbances are considered.
- Deviations of machine frequencies from synchronous frequency are small.
- During short circuit in generator, DC offset may be neglected.
- Network and impedance loads are at steady state. Hence, voltages, currents, and powers can be computed from power flow equation.

In this chapter, steady state stability is assessed for a two-bus system, wherein the direction of power flow is assessed to ascertain the stability of the system. A case study is presented for a sample two-bus system and analysis is done at the end to authenticate the case concerned.

The advantage of the P-δ plane approach is that, here, fault bus and type of fault can be analyzed from other distant buses and lines also, which may not be connected directly with the fault buses. Both symmetrical and unsymmetrical fault analysis can be done using this technique and stability assessment of the connected lines can be done at normal and faulty conditions using this method of analysis in harmonic-infiltrated environment. This is achieved by monitoring harmonic load angle and feeder operating points in P-δ plane. Harmonic P-δ characteristics can be formed and feeder operating points can be determined for healthy and at symmetrical, line to line, line to ground, and double line to round fault in certain buses of the system. Significant changes will be observed in the harmonic operating points (HOPs) due to fault. Also, the change of harmonic load angle due to fault at different frequencies is observed. Based on the observation logic can be made to judge whether the system is healthy or faulty. If there is any fault, then fault bus along with the type of fault is identified. The specific outcome of this method is that, here, fault bus and type of fault can also be assessed from other buses and feeders which may not be connected directly with fault bus by using harmonic voltage angle and power flow monitoring, and logic is developed accordingly. This method can be extended for other multi-bus systems and for identification of other types of faults. Also, steady state stability analysis can be done with the aid of this approach [10].

2.3 Power Equation

Power delivered to load by a synchronous generator depends on generated voltage, load voltage, *i.e.*, the voltage across the load, load angle, *i.e.*, the angle between these two voltages, and reactance of the generating system. Consider a simple single-generator-single-load system as shown in Figure 2.1 [11–13].

Let,

E = *Generated voltage*
V = *load voltage*
δ = *load angle*
δ = *reactance*

Then active power can be mathematically written as follows:

$$P = \frac{EV}{X}\sin\delta \tag{2.1}$$

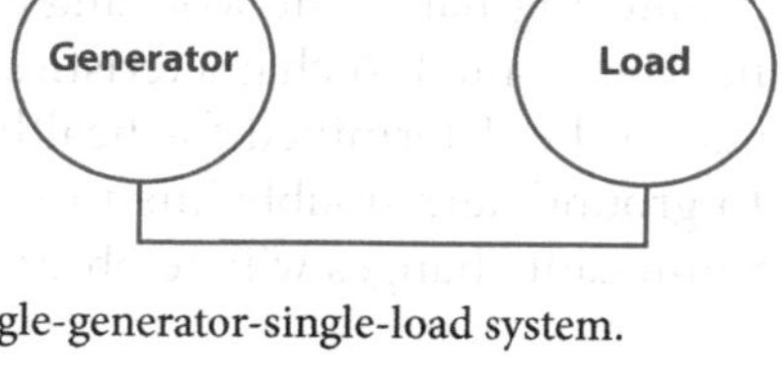

Figure 2.1 A simple single-generator-single-load system.

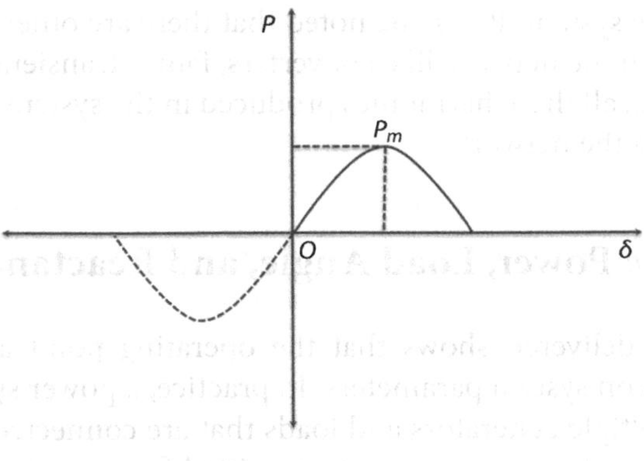

Figure 2.2 Power versus load angle characteristics.

This equation is commonly known as the power equation. The equation shows that power delivered to load varies in the sinusoidal way with load angle. Power versus load angle characteristics is shown in Figure 2.2.

2.4 Maximum Power

Though the power delivered to the system varies with load angle, the maximum power does not depend on load angle; rather it depends on voltages and reactance of the system. From the above equation maximum power can be written as follows:

$$P_{max} = \frac{EV}{X} \tag{2.2}$$

The above expression shows that maximum power increase with the increase of generated and load voltage and decreases with the increase of system reactance.

2.5 Nonlinearity and Harmonics

The characteristics of magnetic materials are nonlinear. This aspect brings nonlinearity in all electrical machines consisting of magnetic materials and also causes harmonic generation in the system. The properties of the power system network largely depend on the properties of the machines installed in the system. It may be noted that there are other major sources of harmonics in the network like converters, faults, transients, and nonlinear loads. Thus, all these harmonics produced in the system also introduce nonlinearity in the network.

2.6 Active Power, Load Angle, and Reactance

Active power delivered shows that the operating point and maximum power depend on system parameters. In practice, a power system network consists of multiple generators and loads that are connected by buses and different feeders. The active power transmitted from one bus to another is almost similar to the previously discussed power equation. Let,

V_j = *voltage at bus j*
V_k = *voltage at bus k*
δ_{jk} = *difference between voltage of bus j and bus k*
X = *reactance between bus j and bus k*

Therefore, the active power can be written as follows [14–19]:

$$P = \frac{V_j V_k}{X} \sin \delta_{jk} \tag{2.3}$$

Thus, the power flow of the network and its stability will depend on the reactance of the network. Network reactance can be expressed in matrix form. Let,

$$[X] = reactance\ of\ the\ network$$

The reactance matrix can be derived from the admittance matrix of the network. Let,

$$[Y] = admittance\ of\ the\ network$$

Therefore, the reactance can be expressed in terms of admittance, as follows:

$$[X] = [Y]^{-1} \tag{2.4}$$

For a three-bus network, the admittance matrix looks like:

$$[Y]_{3\times3} = \begin{bmatrix} Y_{11} & Y_{12} & Y_{13} \\ Y_{21} & Y_{22} & Y_{23} \\ Y_{31} & Y_{32} & Y_{33} \end{bmatrix} \tag{2.5}$$

In generalized form, the admittance matrix of an n-bus system can be written as follows:

$$[Y]_{n\times n} = \begin{bmatrix} Y_{11} & \dots & Y_{1n} \\ Y_{21} & \dots & Y_{2n} \\ Y_{31} & \dots & Y_{nn} \end{bmatrix} \tag{2.6}$$

2.7 Effects of Harmonics on Stability Model

Reactance can be expressed as follows:

$$X = R + j\omega L \tag{2.7}$$

Neglecting resistance, the reactance will consist of only inductance, as follows:

$$X = j\omega L \tag{2.8}$$

Therefore, the magnitude of the reactance will involve angular frequency and inductance, as follows:

$$|X| = |j\omega L| = \omega L \tag{2.9}$$

Similarly, admittance can be written by inversing reactance, as follows:

$$Y = \frac{1}{X} = \frac{1}{j\omega L} \tag{2.10}$$

Therefore, the magnitude of the admittance can be written as follows:

$$|Y| = \left|\frac{1}{j\omega L}\right| = \frac{1}{\omega L} \tag{2.11}$$

2.7.1 Harmonic Reactance

Reactance is a frequency-dependent term.

$$|X| = |j\omega L| = \omega L = 2\pi f\ L \tag{2.12}$$

where,

$$\omega = 2\pi f \tag{2.13}$$

The system is expected to operate with 5% (it may vary based on regulation guidelines) deviation at the following frequency:

$$f = 50\ Hz\ or\ 60\ Hz \tag{2.14}$$

Any real system consists of harmonics. Therefore, i[th] order frequency may be expressed as follows:

$$\omega_i = 2\pi f_i \tag{2.15}$$

where,

$$i = 1, 2, 3, 4, 5 \ldots n$$

If any system is free from even harmonics, then

$$i = 1, 3, 5 \ldots n \tag{2.16}$$

However, some even harmonics with very small amplitude may be found. In those cases, second-order harmonics are found dominating (*e.g.*, inrush in transform network). Therefore, considering this,

$$i = 1, 2, 3, 5 \ldots .n \tag{2.17}$$

Now because of semiconductor-based generating units and controlling devices in practice, the network may consist of fractional harmonics. Then,

$$i = fractional \tag{2.18}$$

In the presence of harmonics, the magnitude of reactance can be written as follows:

$$|X_i| = |j\omega_i L| = \omega_i L \tag{2.19}$$

2.7.2 Harmonic Power Equation and Harmonic Maximum Power

For each harmonic frequency, active power depends on harmonic voltages. Let,

V_{j_i} = *voltage at bus j corresponding to frequency component of order i*

V_{k_i} = *voltage at bus k corresponding to frequency component of order i*

δ_{jk_i} = *difference between voltage at bus j and bus k corresponding to frequency component of order i*

X_{jk_i} = *reactance between bus j and bus k corresponding to frequency component of order i*

Considering i[th] order frequency component, the power equation can be written as follows:

$$P_{jk_i} = \frac{V_{j_i} V_{k_i}}{X_{jk_i}} \sin \delta_{jk_i} \tag{2.20}$$

Therefore, harmonic maximum power can be written as follows:

$$P_{jk_{i\,max}} = \frac{V_{j_i} V_{k_i}}{X_{jk_i}} \tag{2.21}$$

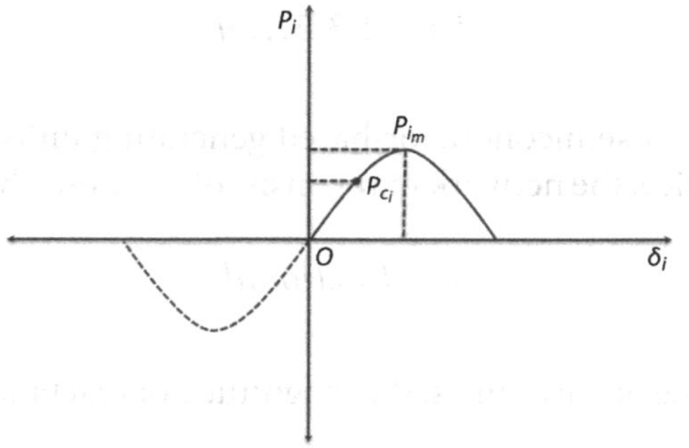

Figure 2.3 Harmonic power, operating point, and corresponding maximum power.

Harmonic power, operating point, and corresponding maximum power are shown in Figure 2.3.

For fundamental components,

$$i = 1 \tag{2.22}$$

Therefore, a fundamental component of maximum power can be written as follows:

$$P_{jk_1\,max} = \frac{V_{j_1} V_{k_1}}{X_{jk_1}} \tag{2.23}$$

Fundamental power, operating point, and corresponding maximum power are shown in Figure 2.4.

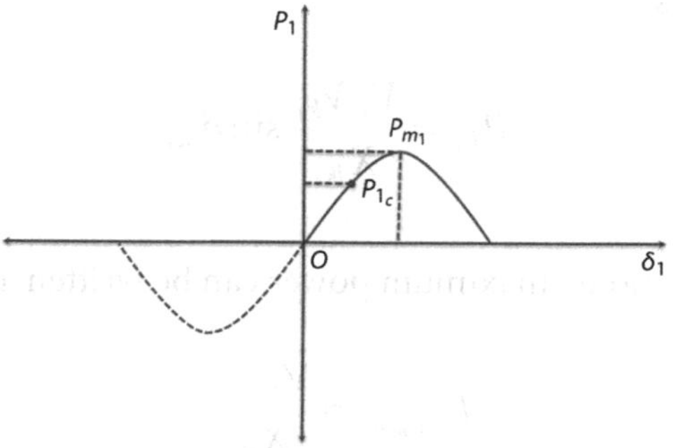

Figure 2.4 Fundamental power, operating point, and corresponding maximum power.

2.8 Harmonic Operating Point

2.8.1 Harmonic Power Versus Load Angle Characteristics

Power versus load angle characteristics is greatly influenced by maximum power. Normally, maximum power or peak power corresponds to the power contributed by fundamental components of voltages and respective reactance of the system. But this maximum power varies with the change of frequency. As frequency of harmonic components differ from that of fundamental, maximum power corresponding to different harmonic components are different and significantly less (most of the cases neglected for power flow analysis and trading aspects) than that of fundamental. Thus, harmonic power versus load angle characteristics refers to the P-power angle characteristics that involve all parameters like voltages, reactance, and power angle having harmonic frequencies.

2.8.2 Harmonic Operating Point

Harmonic operating point (HOP) refers to the operating point concerning active power versus load angle characteristics for a particular frequency. Harmonic operating points are normally seen to vary with frequency.

2.8.3 Does HOP Hamper Overall Stability?

As harmonics' magnitudes are comparatively very small concerning the magnitude of fundamental, harmonic stability does not normally hamper overall stability. Thus, system stability is governed by fundamental components. But in a very small network like nanogrids and picogrids, HOP may influence system stability.

2.8.4 Importance of HOP

Monitoring and study of HOP have the following importance:

- HOPs carry other system information that may be important.
- Harmonics often carry an indication of faults. Therefore, monitoring of HOP along with fundamental operating points can help in fault prediction.
- Performance of control units can be monitored by observing HOPs.

2.8.5 Steps for Determination of HOP

Before determining HOPs, system parameters are to be determined corresponding to harmonic frequencies. Normally, admittance matrix of a particular power system network is determined by changing the frequency dependent matrix elements using harmonic frequencies. Then, using harmonic admittance matrix, harmonic bus voltages are determined. Then, harmonic power angles are determined by the difference of corresponding bus voltages. These angles are used to determine HOPs.

2.9 Case Studies

A case study is presented here to show how HOP changes with frequency. Two-bus network model was considered for harmonic stability study, as shown in Figure 2.5. A power source of 33 kV was applied at sending end. At receiving end, it delivered power to a three-phase load of 300 kW at 33 kV. Harmonic voltages and corresponding interconnecting line reactance were measured, as presented in Table 2.1. Operating load angles for different frequency

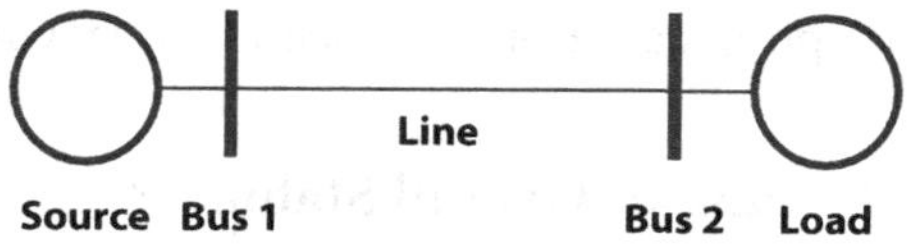

Figure 2.5 Two-bus network model used for harmonic stability study.

Table 2.1 Harmonic voltages and corresponding interconnecting line reactance.

Harmonics	f_iHZ	V_1	V_2	X
Sub harmonics	$f_{0.2}$(10Hz)	7.59	2.33	0.08329164
Do	$f_{0.4}$(20Hz)	24.57	2.82	0.16658328
Do	$f_{0.6}$(30Hz)	40.17	3.42	0.24987492
Do	$f_{0.9}$(40Hz)	86.48	5.63	0.33316656
Do	$f_{1.4}$(70Hz)	67.61	2.49	0.58304148
Do	$f_{1.6}$(80Hz)	42.62	1.30	0.66633312
Do	$f_{1.9}$(90Hz)	37.0	1.06	0.74962476
Integer harmonics	f_3(150Hz)	16.99	0.28	1.2493746

components were calculated, as presented in Table 2.2. Harmonic components of bus voltages excluding fundamental component are shown in Figure 2.6. Load angles for different harmonic frequencies are shown in Figure 2.7.

The same study was carried out for a three-phase load of 150 kW at 33 kV. The harmonics voltages and corresponding reactance are shown in Table 2.3. Corresponding harmonic components of bus voltages excluding fundamental component are shown in Figure 2.8. Operating load angles for different frequency components were calculated, as presented in Table 2.4. Load angles for different harmonic frequencies are shown in Figure 2.9.

Table 2.2 Operating load angles for different frequency components.

Harmonics	f_i**HZ**	$\Delta = \delta_2 - \delta_1$
Fundamental	f_1(50 Hz)	87
Sub harmonics	$f_{0.2}$(10 Hz)	85.5
Do	$f_{0.4}$(20 Hz)	88.7
Do	$f_{0.6}$(30 Hz)	93.7
Do	$f_{0.8}$(40 Hz)	84.6
Do	$f_{1.4}$(70 Hz)	83.4
Do	$f_{1.6}$(80 Hz)	84.4
Do	$f_{1.8}$(90 Hz)	84.6
Integer harmonics	f_3(150 Hz)	79.3

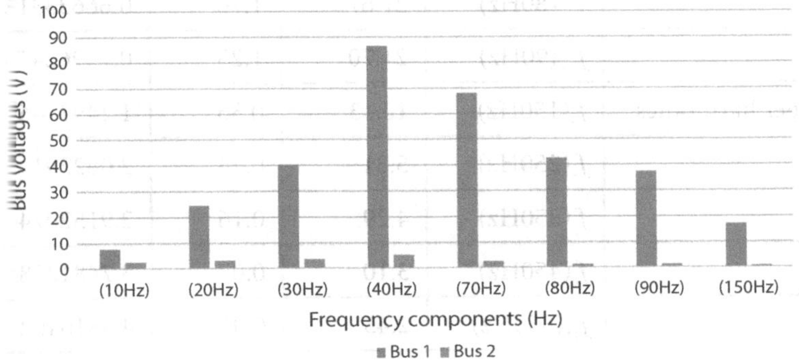

Figure 2.6 Harmonic components of bus voltages excluding fundamental component for 300-kW load.

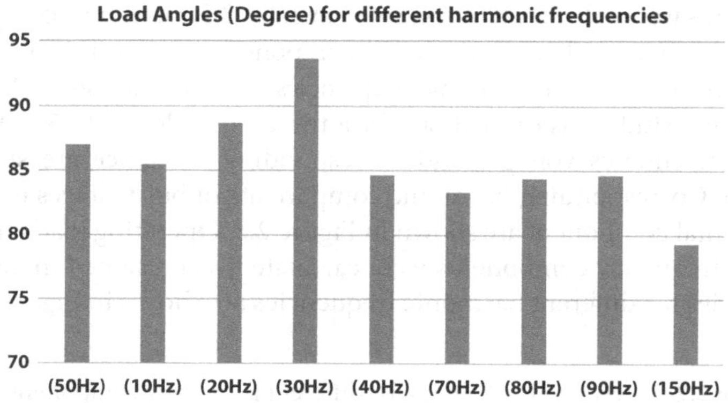

Figure 2.7 Load angles for different harmonic frequencies.

Table 2.3 Harmonic voltages and corresponding interconnecting line reactance.

Harmonics	f_i**HZ**	V_1	V_2	X
Sub harmonics	$f_{0.2}$(10Hz)	5.63	2.84	0.08329164
Do	$f_{0.4}$(20Hz)	12.92	3.25	0.16658328
Do	$f_{0.6}$(30Hz)	25.97	4.32	0.24987492
Do	$f_{0.8}$(40Hz)	52.68	6.82	0.33316656
Inter harmonics	$f_{1.2}$(60Hz)	81.64	7.01	0.49974984
Do	$f_{1.4}$(70Hz)	38.81	2.79	0.58304148
Do	$f_{1.6}$(80Hz)	27.67	1.76	0.66633312
Do	$f_{1.8}$(90Hz)	21.70	1.23	0.74962476
Integer harmonics	f_3(150Hz)	10.13	0.36	1.2493746
Do	f_5(250Hz)	5.31	0.24	2.082291
Do	f_7(350Hz)	4.29	0.16	2.9152074
Do	f_9(450Hz)	3.10	0.05	3.7481238
Do	f_{11}(550Hz)	2.45	0.07	4.5810402

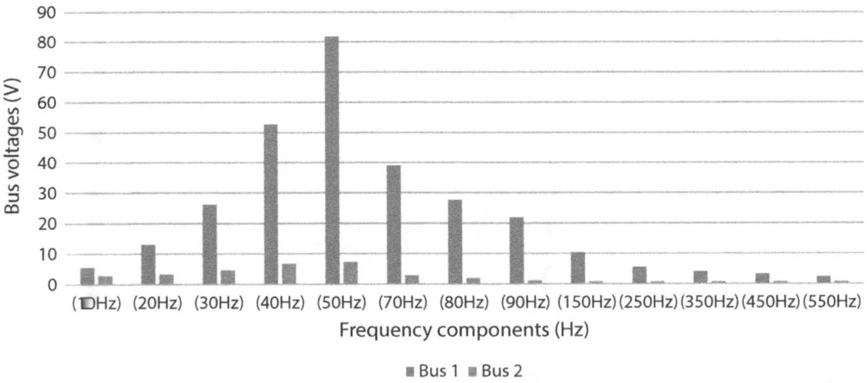

Figure 2.8 Harmonic components of bus voltages excluding fundamental component for 150-kW load.

Table 2.4 Operating load angles for different frequency components.

Harmonics	f_i**HZ**	$\delta = \delta_2 - \delta_1$
Fundamental	f_1(50Hz)	84.3
Sub harmonics	$f_{0.2}$(10Hz)	92.4
Do	$f_{0.4}$(20Hz)	88.7
Do	$f_{0.6}$(30Hz)	88.2
Do	$f_{0.8}$(40Hz)	86.3
Inter harmonics	$f_{1.2}$(60Hz)	82.7
Do	$f_{1.4}$(70Hz)	82
Do	$f_{1.6}$(80Hz)	80.6
Do	$f_{1.8}$(90Hz)	79.3
Integer harmonics	f_3(150Hz)	72.2
Do	f_5(250Hz)	-263
Do	f_7(350Hz)	-10.9
Do	f_9(450Hz)	42.5
Do	f_{11}(550Hz)	65.1

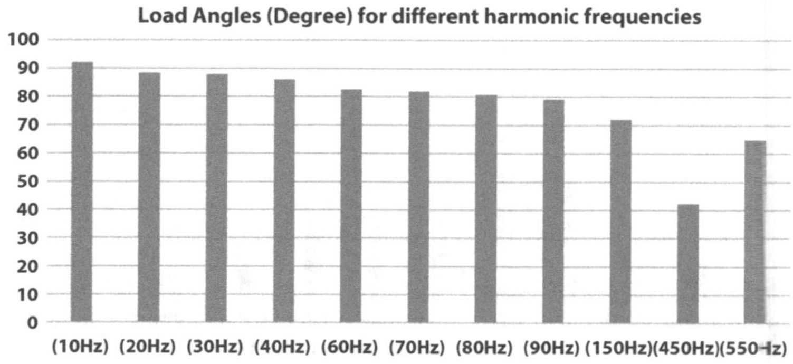

Figure 2.9 Load angles for different harmonic frequencies.

Observation:

The major observations from the case study are as follows:

- Power or load angles for harmonic components are not the same.
- Tables 2.3 and 2.4 show that load angles considerably change with the change of frequency component.
- Sign (+ or -) of power angles are not the same for all harmonic components.
- Power angle features varies with load.
- Further study shows at harmonic power, angle varies significantly during load, thus, these features may be useful for fault diagnosis.

2.10 Conclusions

Different issues of PSS concerning harmonic power flow in the network were presented in this chapter. At first, system stability was defined. Considering gradual load and large disturbances, power system stability was classified. The power equation needed for stability study was expressed. Different useful parameters were pointed out that are important for stability consideration. Then, a detailed discussion was made on how harmonics influence system reactance, operating points that are important for stability analysis. Then the case study results were provided, which show that operating points of different frequency components are different. As the magnitudes of harmonic components are significantly less than that of the fundamental frequency, normally, they do not have a great influence on the overall system stability; however, the study of HOPs is interesting as they can give other important information about the condition of the system.

References

1. C. H. Joe, J. Juan, G. Sanchez, "Flexible AC Transmission Systems," in *Power System Modeling, Computation, and Control*, IEEE, 2020, pp.421-485, DOI: 10.1002/9781119546924.ch14.
2. J. L. R. Torres, F. G. Longatt, "Implementation of a Real Phasor Based Vulnerability Assessment and Control Scheme: The Ecuadorian WAMPAC System," in *Dynamic Vulnerability Assessment and Intelligent Control: For Sustainable Power Systems*, IEEE, 2018, pp.389-411, DOI: 10.1002/9781119214984.ch18.
3. D. Xu, F. Blaabjerg, C. Wenjie, N. Zhu, "Grid Codes for Wind Power Generation Systems," in *Advanced Control of Doubly Fed Induction Generator for Wind Power Systems*, IEEE, 2018, pp.43-63, DOI: 10.1002/9781119172093.ch3.
4. D. Xu, F. Blaabjerg, W. Chen, N. Zhu, "Resonant Control of DFIG Under Grid Voltage Harmonics Distortion," in *Advanced Control of Doubly Fed Induction Generator for Wind Power Systems*, IEEE, 2018, pp.195-235, DOI: 10.1002/9781119172093.ch8.
5. H. Farhangi, G. Joos, "Analysis and Studies Using Recommended Models," in *Microgrid Planning and Design: A Concise Guide*, IEEE, 2019, pp.57-63, DOI: 10.1002/9781119453550.ch4.
6. A. A. Sallam, O. P. Malik, "Distributed Generation," in *Electric Distribution Systems*, IEEE, 2019, pp.487-533, DOI: 10.1002/9781119509332.ch18.
7. S. Chattopadhyay, "*Nanogrids and Picogrods and their integration with electric vehicles*", IET, London, ISBN: 978-1-83953-482-9, 2022.
8. S. Chattopadhyay, A Das, "*Overhead Electric Power Lines: Theory and practice*", IET, London, ISBN: 9781839533112, 2021.
9. S. Chattopadhyay, M Mitra, S Sengupta, "*Electric Power Quality*", Springer, Netherland, ISBN: 978-94-007-0635-4, 2011.
10. D. Kar Ray (2019). "*Harmonic Power Flow Analysis and its Applications in Electrical Systems*", Ph. D. Thesis, UCSTA, CU, Kolkata, India.
11. T. Zabaiou, L. A. Dessaint, I. Kamwa, "Preventive Control Approach for Voltage Stability Improvement using Voltage Stability Constrained Optimal Power Flow based on Static Line Voltage Stability Indices", *IET Gen. Trans. Distr.*, Vol. 8 (5), pp. 924-934, doi:10.1049/iet-gtd. 2013. 0724, 2014.
12. P. M. Anderson, A. A. Fouad, "*Power System Control and Stability*", IEEE Press, ISBN (Print): 978-0-471-23862-1, 2003.
13. P. W. Sauer, M. A. Pai, "*Power System Dynamics and Stability*", Stipes Publishing Co., ISBN-10:1588746739, ISBN-13:978-1588746733, 2016.

14. O. P. Rahi, A. K. Yadav, H. Malik, A. Azeem, B. Kumar, "*Power System Voltage Stability Assessment through Artificial Neural Network*", ELSEVIER, Vol. 30, pp. 53-60, doi: 10.1016/j. proeng. 2012. 01. 833, 2012.
15. G. A. Mahmoud, "Voltage Stability Analysis of Radial Distribution Networks using Catastrophe Theory", *IET Gen. Trans. Distr.*, Vol. 6 (7), pp. 612-618, doi: 10. 1049/iet-gtd.2011. 0530, 2012.
16. P. Kundur, "*Power System Stability and Control*", NewYork: McGraw-Hill, 2006.
17. N. R. Bujal, A. E. Hasan, M. Sulaiman, "*Analysis of Voltage Stability Problems in Power System*", in Proc. ICE2T, Malaysia, pp. -278-283, doi: 10. 1109/ICE2T. IEEE, 2014.
18. P. W. Sauer, M. A. Pai, "*Power System Dynamics and Stability*", Stipes Publishing Co., ISBN-10:1588746739, ISBN-13:978-1588746733, 2003.
19. O. L. Elgerd, "*Electric Energy Systems-Theory an Introduction*", McGraw-Hill, ISBN 0070191316, 9780070192317, 1982.

3

Comparative Study of Different Existing Standard Microgrid Networks

Sagnik Datta[1]*, Aveek Chattopadhyaya[2], Surajit Chattopadhyay[3] and Arabinda Das[4]

[1]*Electrical Engineering Department, S.K.F.G.I., Mankundu, Hooghly, India*
[2]*Electrical Engineering Department, GNIT, Panihati, Kolkata, India*
[3]*Electrical Engineering Department, G.K.C. Institute of Engineering and Technology, Malda, India*
[4]*Electrical Engineering Department, Jadavpur University, Kolkata, India*

Abstract
Reliability has become one of the most important properties of a power distribution network in modern era. Nowadays, a distribution network faces threats of cyberattacks on it along with weather-related events, such as storms and floods. Microgrids gained extreme popularity for enhancing overall reliability and resilience of conventional power networks. It consists of a set of distributed energy sources and loads present in a small region, which acts as a separate controllable set-up in relation to the main grid. However, widespread introduction of microgrid introduces some new technical challenges. Unlike traditional power networks, generators, energy storage facilities, and loads are present at all the levels. Thus, generation, distribution, and consumption can occur at any level, making the direction of power flow very much uncertain. Deep understanding of a microgrid operation is still at its infancy. In order to develop a better understanding, standardized forms of microgrid networks are of paramount importance. This chapter showcases three most widely used standardized networks: IEEE 14 bus microgrid network, IEEE 9 bus microgrid network, and IEC 61850-7-420 standard microgrid network.

Keywords: DC microgrid, AC microgrid, hybrid microgrid, energy storage, energy resource, power distribution, protection schemes

**Corresponding author*: sagnik.ee@gmail.com

Tamal Roy, Suman Lata Tripathi and Souvik Ganguli (eds.) Linear and Nonlinear System Modeling, (33–48)

3.1 Introduction

In recent times, reliability, *i.e.*, the ability to stay 'operational in the event of adversity' is becoming one of the most desirable properties of a power distribution network. The word 'adversity' refers to weather-related events, such as storms and floods, and also physical/cyberattacks on power network. Evolutions in the area of distributed energy resource (DER) and energy storage system are making microgrids extremely popular for enhancing overall reliability and resilience of conventional power networks. Microgrid consists of a set of distributed energy sources and loads present in a small area, which acts as a separate controllable set-up with respect to the main grid, having a distinct electrical borderline. It is capable of working either connected with the traditional power grid or on its own as an island. However, widespread introduction of microgrid will bring forth its own sets of technical challenges. The main challenges deal with the significant impact on existing networks created due to the deployment of a number of DERs and design of proper protection schemes which can operate satisfactorily in the much lower voltage and current levels present in the microgrid environment. Unlike traditional power networks, generators, energy storage facilities, and loads are present at every level of a microgrid network. Generator, distributor, and also consumer can stay at any level of a microgrid network and the direction of power flow is also uncertain. Complete understanding of a microgrid operation is still at a very early phase. Thus, to facilitate a secure operation in such a multifaceted environment, where so many different parameters and variables are to be monitored, controlled and communicated, extensive research work is needed. In order to do so, standardized forms of microgrid networks are of paramount importance. This chapter showcases three most widely used standardized networks: IEEE 14 bus microgrid network, IEEE 9 bus microgrid network, and IEC 61850-7-420 standard microgrid network.

3.2 Classification of Microgrid Networks

Microgrid networks can be classified in three broad categories, depending upon the DER units and the type of loads present. These are DC microgrid, AC microgrid, and hybrid microgrid.

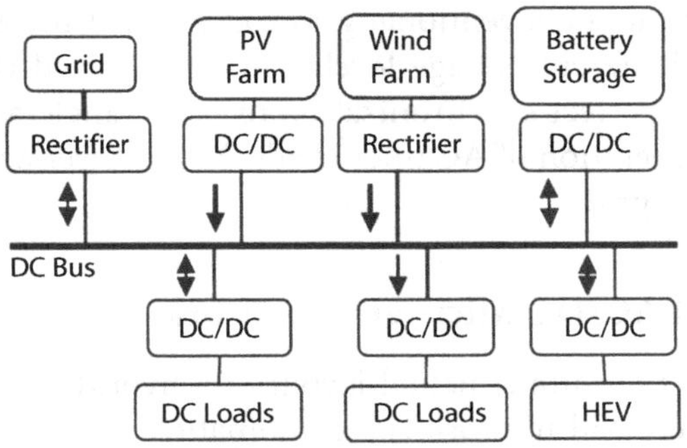

Figure 3.1 Basic structure of a DC microgrid network.

3.2.1 DC Microgrid Network

DC microgrids have the lowest energy loss in comparison the other kinds of microgrid systems. Figure 3.1 shows the basic layout of a DC microgrid which can operate at different direct current (DC) voltages. The control operation is relatively easier due to the absence of reactive power and only the amplitude of voltage needs to be regulated for connecting DC sources with the bus. The absence of energy conversion stages and no need for phase and frequency synchronization are the two main advantages of DC microgrid networks. Usage is limited only to applications requiring low voltage levels.

3.2.2 AC Microgrid Network

Figure 3.2 shows the basic form of an AC microgrid network. The main advantage of this type of microgrid is that local loads can draw power

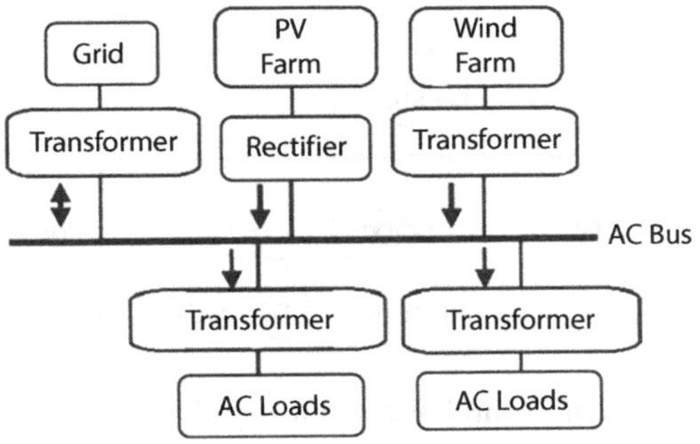

Figure 3.2 Basic structure of an AC microgrid network.

from it and the main conventional grid can also get the excess power of the microgrid. Different voltage levels can be achieved through the use of transformers. Perfect synchronization of phase angle and frequency is needed for the operation of AC microgrids. High power applications need this type of microgrids.

3.2.3 Hybrid AC/DC Microgrid Network

Figure 3.3 shows the fundamental layout of a hybrid AC/DC microgrid. This type of microgrid network is a combination of AC and DC microgrids. So, it has two distinct portions. Direct connection between the AC network and the conventional grid is present. An interlinking converter joins the DC section of the microgrid network with the conventional power grid. These types of networks are designed by considering the flow of power, nature of loads, and cost. This type of microgrid network can serve both AC and DC loads. All the existing standard forms of microgrids are of this type.

3.3 Modes of Operation

IEEE 2030.7-2017 describes the different operating modes for a microgrid network in a comprehensive way. This standard summarizes microgrid operation into steady state (SS) operating mode and transitions (T) of different kinds, as depicted in Figure 3.4.

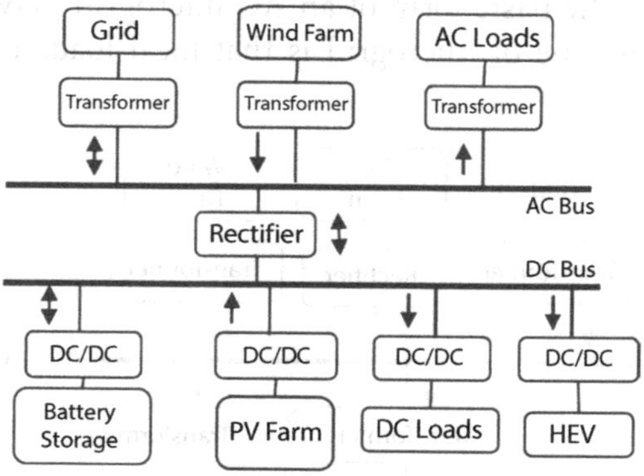

Figure 3.3 Basic structure of a hybrid AC/DC microgrid network.

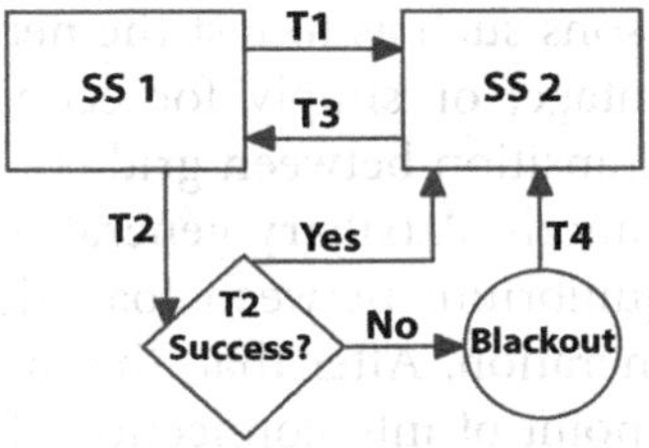

Figure 3.1 Microgrid operating modes and control flow diagram.

The operating modes according to IEEE 2030.7-2017 are as follows:

- SS1: steady state (grid connected): here, the network is in connection with traditional grid. It is capable of providing services to the main grid, such as peak shaving, valley filling, and reactive power control. The microgrid controller provides a single point of communication for availing all of the services.
- SS2: steady state (island): here, disconnection between the microgrid and the traditional power network will be achieved. So, the microgrid network operates on its own (formally known as "islanding"). A continuous balance needs to be present between loads and local generation and energy storage facilities. If the microgrid has enough generation capacity then it can be operated in this mode indefinitely. Island mode demands different settings for the protective devices than that needed for grid-connected condition, due to much reduced levels of faulty current and possible alterations in power flow directions. Load management is crucial in island mode for achieving operating stability in case generation is not enough for the net load demand, or if the energy storage systems do not have enough reserve to meet power demand for a specific time frame. Microgrids which only have solar or wind energy sources generally need systems to store energy, for getting some buffer in between different sources and the loads. It also requires the capability to cut down generation, when the storage gets full and generation becomes more than load demand, in order to prevent high frequency or voltage-related problems.
- Transition T1: from grid connected to steady state island (planned): microgrid may be operated in island mode even when the traditional power grid is available. It can be done

for several reasons such as to test the network for a scheduled power outage, or simply for economic reasons. T1 demands the transition between grid connected and island mode to be smooth. Auxiliary generation gets initiated to achieve an equilibrium between load demand and available power generation. After that, the microgrid is disconnected at the 'point of interconnection (POI).' At the same time, local generation should also switch from 'grid follower mode' to 'grid former mode'. Protective devices generally operating in a particular direction have to be bidirectional at island mode. It is extremely important that their settings are tweaked accordingly.

- Transition T2: From grid connected to steady state island (unplanned): here, the operation takes place if in the event of fault occurring in main network, the microgrid has the ability for disconnecting itself and transit into island mode without interruption of power supply to the loads with the help of local generations. T2 is somewhat easier in case of small dedicated microgrid networks but difficult for large microgrid networks. Sometimes, the critical loads are disconnected during this process until additional power sources are brought online. Similar to the planned island, protective equipment require adjustments.
- Transition T3: steady state island reconnection with grids: here, changeover happens from the island mode of operation to grid connected through successful reconnection. Both frequency and phase angle of the 'grid forming' generator present in microgrid network must be matched with that of the traditional power grid to obtain proper synchronization. After the reconnection is complete, the 'grid-forming' generator gets switched to the 'grid following'.
- Transition T4: black start to steady state island: in this mode of operation, a microgrid is restarted in island mode after experiencing black out. T4 needs the microgrid network to be detached from the main grid at the POI. If the microgrid controller fails to deal with an unexpected power outage of the main power grid and initiate stable transition (T2) to island mode then T4 is experienced. T4 can also happen if there exists insufficient generation or energy storage reserve for continuation of supply to loads, and consequently, it has to be shut down.

3.4 General Equipment of a Microgrid Network

Identification of the energy sources, storage for energy, management of load, and automation enabled, distribution-related equipment, which are present in the microgrid, is of paramount importance for understanding the operation of the microgrid controller.

- Energy sources: energy sources present in microgrid networks are generally photovoltaic systems (PV), diesel generators, fuel cells, and wind generators. Nowadays, behind-the-meter (BTM) sources like commercial or residential PV systems are also being utilized. One of the active energy sources act as the 'grid-forming' generator. It sets the operating frequency and voltage level of the network. A conventional alternator or an inverter along with battery set-up is used as the 'grid-forming' generator. Solar systems and wind generators generally do not play the role of 'grid-forming' generator due to their highly variable nature.
- Energy storage: energy storage facilities act like a bridge between generation and load management. Battery systems can also serve as energy sources when there is a provision of recharging them from other sources. Thermal-energy storage, *e.g.*, controlled water-heater or storage of ice for commercialized cooling purpose, are also useful in enhancing grid stability, though they are not active contributors of electric power. Table 3.1 shows the classification of systems for energy storage based on their working principles.
- Load management: it is the key component of the balanced operation of a microgrid network because proper load management ensures a balance between generations and the load demand. Load management can be of various types such as techniques needed to respond against utility demand, remotely disconnecting meters at domestic and commercial (small) facilities, active load-disconnection facility at commercial (large) units and sectionalization of feeders for the purpose of isolating entire groups of users based on their location. Information must be fed to the microgrid controller regarding the loads which are present at various positions inside the network. Priority wise plan for the most critical load should also be maintained, in case of an event of limited generation is faced.

Table 3.1 Classification of energy storage systems on the basis of working principle.

Sl. no.	Types of energy storage	Examples
1.	Mechanical	• Flywheel • Compressed air • Hydro pump
2.	Electrical	• Super conducting magnetic energy storage • Capacitor bank
3.	Electrochemical	• Secondary battery (lead acid, NiCd, NaS) • Flow battery (redox flow, hybrid flow) • Super capacitor
4.	Chemical	• Fuel cell
5.	Thermal systems	• Heat storage

- Automated distribution system equipment: two purposes are generally served by automated distribution system equipment present in a microgrid network. Firstly, they perform micro-level load management, such as initiating 'rotating blackouts' in case a shortage in generation is faced. Secondly, they also play the role of sensing as well as protective equipment. The electrical characteristics vary considerably, between island mode and also grid-connected mode of working because flow of current in grid-connected mode is always unidirectional—from substation to load, whereas, in steady state island (SS2) mode of operation, flow of current can be bidirectional depending upon the energy sources available at a given time. Hence, the protective devices should possess the ability to handle the bidirectional current flow in case the microgrid network experiences a changeover between grid connected and island mode.
- Communications and cyber security issues: for proper operation of a microgrid network, microgrid controllers need to communicate with a number of devices such as inverters present in PV systems to residential thermostats. This communication system poses the greatest challenge in design and operation of a microgrid network mainly because of a great variety of communication technologies (*e.g.*, fiber, PLC,

cellular, etc.) and also of standards and protocols (*e.g.*, Modbus, DNP3, IEC 61850, etc). The protocols generally employed in the network are also very much different than the consumer friendly BTM arrangement. Controllers of microgrids are either self-reliant or hierarchical in nature. In both cases, interoperability is very much needed. Cyber security is also very closely related to communications and interoperability. Devices should possess the abilities to decipher out a valid status request or control from one that is not. It should also ensure the security of communication signals and decryption only by the proper controller. Hence, cyber security needs to be ingrained with software design at the earliest stage.

3.5 Basic Control Structure of Microgrid Network

The basic control structure for microgrid operation generally consists of three stages or layers.

- Primary control: output voltage regulation through inverters and sharing of load among the inverters is performed at primary level of control. This level of control is required for maintenance of economy and enhancing the overall reliability.
- Secondary control: deviations in voltage waveforms as well as frequency caused at primary control level is rectified with the help of secondary control. Load demand forecasting and adjustment of output from the DERs are performed.
- Tertiary control: exchange of power between different consumers is managed at this control level. It is utilized to ensure the reliability of economic-emission dispatch. Optimization-based control techniques are utilized to implement this control.

3.6 Existing Standard Models

The behavior of a microgrid network, regardless of the mode in which it is operating, is extremely dynamic. Connection or disconnection of a DER or a load can take place at an instant. In case of such a change in the network, suitable changes need to be made to the power generation, load-sharing, control, and protection schemes. Despite the recent popularity of microgrid, it is still in very early phases of understanding. Hence, to understand and study the myriad technical challenges present in operation, control, and protection of microgrid networks, standardized forms are very helpful. These standard networks can also be used for research and development of new theories for microgrid environment. Three of the most popular and widely utilized standard microgrid networks are IEEE standard 14 bus microgrid network, IEEE 9 bus microgrid network, and IEC 61850-7-420 standard microgrid network. Detailed discussion regarding these networks is provided in the following section.

3.6.1 IEEE 14 Bus Microgrid Network

It is one of the earliest developed standard model of microgrid network derived from the conventional IEEE 14 bus network. IEEE standard 14 bus microgrid is shown in Figure 3.5 [1]. There are four different types of sources in this microgrid network. Bus G_{13} has the bulk generator connected to it. At G_{12}, PV system is present. Diesel generator 1 (DG1) and diesel generator 2 (DG2) are connected at G_8 at G_3, respectively. Loads are of four different types. Bus B_{10} has nonlinear bulk load, B_6 has furnace, B_1

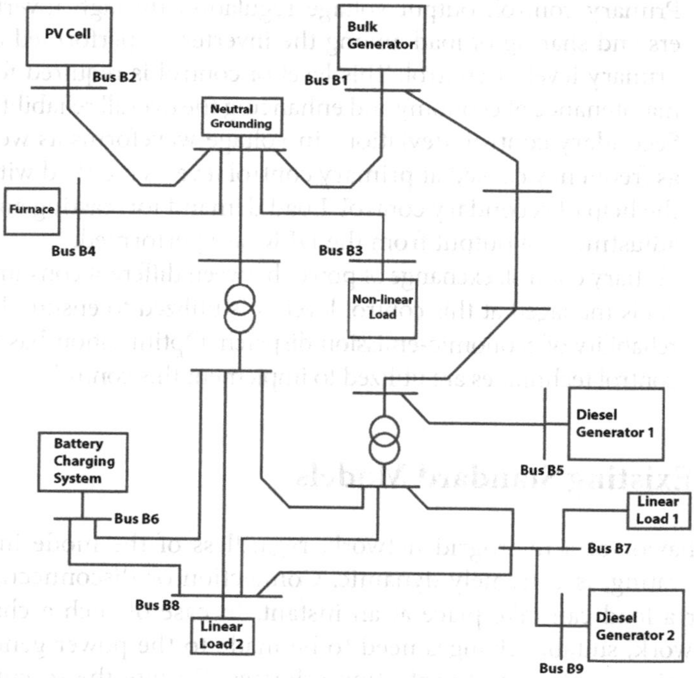

Figure 3.5 IEEE Standard 14 bus microgrid network.

has an arrangement for battery charging, B_2 has linear load, and B_3 has linear load 1 connected to it. The length of transmission lines connecting one bus to its adjacent bus is 1 km. Operating voltage and frequency is kept at 400 V (line to line) and 50 Hz, respectively. Bulk generator bus is considered as the swing bus. Bulk generator is rated at 100 MW, PV system has a capacity of 40 MW, and both the diesel generators are rated at 50 MW.

3.6.2 IEEE 9 Bus Microgrid Network

This network is derived from the conventional IEEE 9 bus network. It has been modified into a microgrid network by reducing the line lengths and strategically introducing various DERs [2]. Alternator G1 is taken as the reference generator and bus B1 is considered as the swing bus. Three alternators present in this network are of three different types. The hydro-alternator is connected to bus B1, gas-turbine is at B2, and the coal-fired type alternator is present at B3. The capacity of the hydro-alternator, gas turbine, and the coal-fired type alternator is 250 MW, 255 MW and 230 MW, respectively. The automatic voltage regulator that is present with each of these alternators has a turbine. Solar PV systems are connected at buses 5 and 9, doubly-fed induction generator-based wind farms are at buses 6 and 8. These sources are connected to the microgrid through a transformer with a rating of 0.4/230 kV. Power rating of the PV farm A and B are 69 MW and 49 MW, respectively. The wind farm has a power rating of 50 MW. An energy storage system (BESS) is connected with bus 1. IEEE 9 bus microgrid network is shown in Figure 3.6.

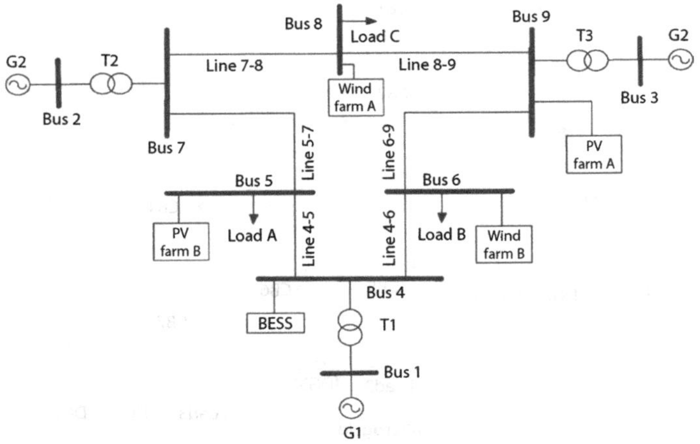

Figure 3.6 IEEE 9 bus microgrid network.

3.6.3 IEC 61850-7-420 Standard Microgrid Network

In IEC 61850-7-420 standard network, all branches contain energy source and load, and the main feature of this network is that the structure can be altered as per requirement network with the help of combinations of relays [3]. The power flow, load sharing, stability parameters, and protection schemes need to be adjusted accordingly. In this network structure, various DGs and loads are connected to and disconnected from the microgrid followed by some minor adjustments. In case CB4 is disconnected, then CB5 has to be closed to keep the integrity of the network. For protection of the microgrid against failures, the line connecting Loads 1 and 2 (under the protection of CB5) forms a loop structure if needed. The whole network structure and the relationship between the microgrid components change significantly. This would require major adjustments on the flow of power, sharing of load, stability parameters, and schemes for protection to ensure satisfactory and safe operation [8–12]. Therefore, a centralized microgrid management system which implements extensive communication with microgrid components is required [13–19]. This management system will monitor DGs, loads, storage devices along with protection equipment, and assign necessary parameters for respective component. The continuous monitoring and adjusting micro-grid parameters on the fly help implement a dynamic management system required by the highly variable nature of the microgrid. Figure 3.7 shows the IEC 61850-7-420 standard microgrid network.

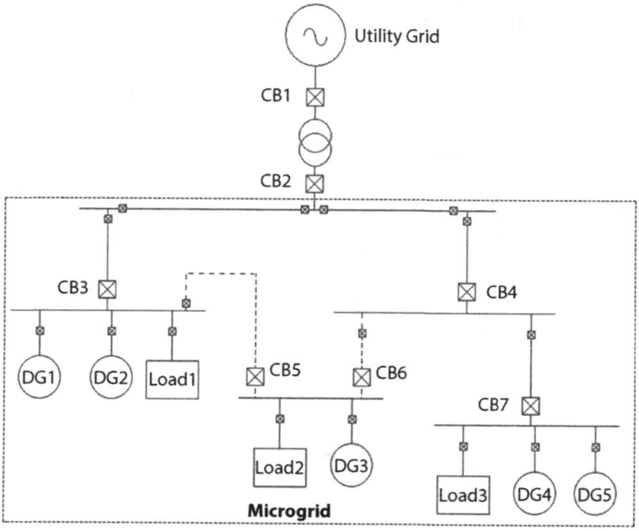

Figure 3.7 IEC 61850-7-420 standard microgrid network.

3.7 Considerations for Designing of Protection Schemes

Two fundamental differences between traditional power grid and microgrid network are that in a microgrid environment, neither the structure of the grid is 'radial' nor is it 'passive' in nature. Microgrids also have much lower levels of voltages and currents in comparison with the traditional power grids. So, the protective equipment and protection schemes employed for protection of traditional grids do not work satisfactorily. Moreover, the direction of power flow in a microgrid network may change depending upon its mode of operation, time of the day, and available sources, which adds additional challenges in designing a protection strategy [20–27].

Aforementioned challenges call for a centralized protection scheme which will be independent of voltage and current levels, direction of current flow, and fault impedance. Some advancement in this regard has been made by considering the statistical nature of harmonic contents in the fault current waveforms [1, 4–7].

3.8 Conclusion

The presence of sources like small PV panels, wind turbines, and diesel generators in a microgrid network grid brings forth marked differences between the structure of a traditional power grid and a microgrid. These radical changes in structure create challenges in operation and protection of a microgrid network which were not known by the power engineers. Traditional load flow control methods have also been rendered ineffective in a microgrid environment along with the popular protection schemes. Hence, rigorous study and research work is needed to address the various issues faced in a microgrid environment. The standardized microgrid networks play an important role in studying the nature of microgrids in different modes of operations as well as devising new operation and protection strategies.

References

1. S. Datta, A. Chattopadhyaya, S. Chattopadhyay & A. Das (2020), "S-Transform Based Kurtosis Analysis for Detection of LG and LL Faults in 14 Bus Microgrid System," *IETE J. Res.*, Vol. 2, no. 1, pp. 10_20.

2. U. Datta, A. Kalam, & J. Shi (2020), "Battery Energy Storage System for Aggregated Inertia-Droop Control and a Novel Frequency Dependent State-of-Charge Recovery", *Energies*, Vol. 13, no. 8, pp. 1_18.
3. T. S. Ustun, C. Ozansoy, & A. Zayegh (2011), "Distributed Energy Resources (DER) object modeling with IEC 61850–7–420," *Australasian Universities Power Engineering Conference (AUPEC 2011)*, Brisbane, Australia, pp. 1_6.
4. M. Mishra, M. Sahani, & P.K.Rout (2017), "An islanding detection algorithm for distributed generation based on Hilbert–Huang transform and extreme learning machine", *Sustain. Energy Grids Netw.*, vol. 9, no. 1, pp. 13_26.
5. M. Manohar, E. Koley, & S. Ghosh (2017), "Reliable protection scheme for PV integrated microgrid using an ensemble classifier approach with real-time validation", *IET Sci. Meas. Technol.*, vol. 12, no. 2, pp. 200_208.
6. S. Datta, S. Chattopadhyay and A. Das (2017), "FFT based Harmonic Assessment of Line to Ground Fault in 14 Bus Microgrid System," *2017 International Conference on Modelling and Simulation in Science, Technology and Engineering Mathematics (MS-17)*, Kolkata, India, pp. 73_86.
7. N. Mukherjee, A. Chattopadhyaya, S. Chattopadhyay, & S. Sengupta (2020), "Discrete-Wavelet-Transform and Stockwell-Transform-Based Statistical Parameters Estimation for Fault Analysis in Grid-Connected Wind Power System", *IEEE Syst. J.*, vol. 14, no. 3, pp. 4320_4328.
8. T. S. Ustun, C. Ozansoy, & A. Zayegh (2012), "Modeling of a Centralized Microgrid Protection System and Distributed Energy Resources According to IEC 61850-7-420", *IEEE Trans. Power Syst.*, Vol. 27, no. 3, pp. 1560_1567.
9. S. Dhar, R. K. Patnaik, & P. K. Dash (2017), "Fault Detection and Location of Photovoltaic Based DC Microgrid Using Differential Protection Strategy", *IEEE Trans. Smart Grid*, vol. 9, no. 5, pp. 4303_4312.
10. T. Ghanbari, & E. Farjah (2013), "Unidirectional Fault Current Limiter: An Efficient Interface between the Microgrid and Main Network", *IEEE Trans. Power Syst.*, vol. 28, no. 2, pp. 1591_1598.
11. T. S. Ustun, C. Ozansoy, & A. Zayegh (2011), "Extending IEC 61850-7-420 for distributed generators with fault current limiters", *IEEE PES Innov. Smart Grid Technol.*, Perth, Australia, pp. 1_8.
12. B. M. Eid, N. A. Rahim, J. Selvaraj, & A. H. El. Khateb (2016), "Control Methods and Objectives for Electronically Coupled Distributed Energy Resources in Microgrids", *IEEE Syst. J.*, Vol. 10, no. 2, pp. 446_452.
13. D. Rebollal, M. Carpintero-Rentería, D. Santos-Martín, & M. Chinchilla (2021), "Microgrid and Distributed Energy Resources Standards and Guidelines Review: Grid Connection and Operation Technical Requirements," *Energies*, Vol. 14, no. 3, pp. 1_24.
14. Y. Usama, & X. Lu (2014), 'Design and implementation of a wavelet analysis-based shunt fault detection and identification module for transmission lines application', *IET Gener. Transm. Distrib.*, Vol. 8, no. 3, pp. 431_441.

15. D. Guillen, M. Paternina, & J. Ortiz-Bejar (2018), 'Fault detection and classification in transmission lines based on a PSD index', *IET Gener. Transm. Distrib.*, Vol. 12, no. 18, pp. 4070_4078.
16. A. Ali, Q. Khan, & A. Hussain (2015), 'Fault modelling and detection in power generation, transmission and distribution systems', *IET Gener. Transm. Distrib.*, Vol. 16, no. 3, pp. 2782_2791.
17. D. Salomonsson, L. Soder, & A. Sannino, 'Protection of Low-Voltage DC Microgrids', *IEEE Trans. Power Deliv.*, Vol. 24, pp. 1045_1053.
18. S. Wen, Y. Wang, & Y. Tang (2019), 'Proactive frequency Control based on Ultra-Short-Term Power Fluctuation Forecasting for High Renewable Penetrated Power Systems', *IET Renew. Power Gener.*, Vol. 13, no. 12, pp. 2166_2173.
19. J. Chen, C. Kuo, & S. Chen (2019), 'DC-side fault detection for photovoltaic energy conversion system using fractional-order dynamic-error-based fuzzy Petri net integrated with intelligent meters', *IET Renew. Power Gener.*, Vol. 10, no. 9, pp. 1318_1327.
20. Y. M. Makwana, & R. B. Bhalja (2017), 'Islanding detection technique based on superimposed components of voltage', *IET Renew. Power Gener.*, Vol. 11, no.11, pp. 1371_1381.
21. M. Manohar, E. Koley, & S. Ghosh (2019), 'Enhancing the reliability of protection scheme for PV integrated microgrid by discriminating between array faults and symmetrical line faults using sparse auto encoder', *IET Renew. Power Gener.*, Vol. 13, no. 2, pp. 308_317.
22. R. Mohanty, U. S. M. Balaji, & A. K. Pradhan (2016), 'An Accurate Non-iterative Fault-Location Technique for Low-Voltage DC Microgrid', *IEEE Trans. Power Deliv.*, Vol. 31, no. 2, pp. 475_481.
23. W. Qiu, Q. Tang, J. Liu, Z. Teng, & W. Yao (2019), 'Power Quality Disturbances Recognition Using Modified S – Transform and Parallel Stack Sparse Auto-encoder', *Electr. Power Syst. Res.*, Vol. 174, no. 1, pp. 1_10.
24. C. Lin, W. Gao, & M. F. Guo (2019), 'Discrete Wavelet Transform Based Triggering Method for Single Phase Earth Fault in Power Distribution Systems', *IEEE Trans. Power Deliv.*, Vol. 34, no. 5, pp. 2058_2068.
25. E. Kumar, B.Singh, D. T. Shahani, A. Chandra, & K. Al-Haddad (2015), 'Recognition of Power Quality Disturbances Using S – Transform Based ANN Classifier and Rule Based Decision Tree', *IEEE Trans. Ind. Appl.*, Vol. 51, no. 2, pp. 1249_1258.
26. M. Ikhide, S. B. Tennakoon, H. Ha, A. L. Griffiths, S. Subramanian, & A. J. Adamczyk (2019), 'A transient and non-unit-based protection technique for DC grids based on the rate-of-change (R-o-C) of the fault induced travelling wave components', *Sustain. Energy Grids Netw.*, Vol. 17, no. 1, pp. 100_11.

4

Application of Active Power Filter in the Hybrid Power System to Regulate the Grid Voltage

Sarita Samal[1]*, Rudranarayan Dash[1], Arjyadhara Pradhan[1] and Prasanta Kumar Barik[2]

[1]School of EE, KIIT DU, Bhubaneshwar, Odisha, India
[2]Dept. of Mechanical &Electrical Engineering, CAET, OUAT, Bhubaneshwar, Odisha, India

Abstract

This work outlines a strategy for integrating solar photovoltaic (SPV) and wind energy system (WES) connections. Both rely on storage batteries for energy delivery, which are one of the distribution system's key advantages. This technique is used to control the voltage at the point of common coupling (PCC) while the SPV system is not producing any power. Using the SPV and batteries as storage energy, the overall system was able to achieve bidirectional grid voltage regulation. During voltage disruptions like sag, swell, and interruption, the bidirectional inverter functions as a normal active power filter, regulating the grid voltage. When the SPV system is using a battery charger, particularly at night time to charge the batteries, it will behave like a conventional switch, and energy will be stored. The proposed technique established the idea of indirect feeder link pathways to regulate voltage regulation, in which as the voltage level rises, a large amount of reverse power flow from WES is automatically regulated by employing the SPV system inverter to charge and store the power. Simulink blocks in MATLAB can be used to model the suggested system. The proposed strategy to adjust the feeder voltage by exchanging real active power through the energy storage batteries could be verified by simulation results. Further, the proposed system may be tested with different recent control approaches of active filters for better performance.

Keywords: Active power filter, solar PV, wind energy, voltage sag, swell

**Corresponding author*: ssamalfel@kiit.ac.in

Tamal Roy, Suman Lata Tripathi and Souvik Ganguli (eds.) Linear and Nonlinear System Modeling, (49–66)

4.1 Introduction

The performance of the electrical power generation has greatly deteriorated recently as a consequence of increased power utilization. The use of nonlinear loads in the system results in harmonics, which affect the current and voltage waveforms and reduce the power factor of the supply system [1]. Consequently, nonlinear loads in the system must be regulated for harmonics. Because passive filters are inexpensive and reliable, they have been employed for harmonic reduction in the past, but they have several problems, including resonance and mistuning, which cause their interruption in the network [2, 3]. The passive filters, have the drawbacks of being costly and having large inverter losses [4, 5]. As a consequence, this chapter presents a series active power filter (APF) incorporated with a hybrid power system in conjunction with multiple control approaches to eliminate all of these shortcomings and produce superior harmonic mitigation at a lower overall cost [6, 7]. The authors of References [8] and [9] examined the dynamics of a combined series APF and battery energy storage (BES) combined arrangement to reduce resonance issues during transient conditions. A PI (proportional integration) control was used to drive the voltage source control (VSC). The authors of References [10] and [11] projected a PI-controlled series APF with a combined storage of a BES and a super capacitor to increase voltage stability. In References [12] and [13], an enhanced modular multilevel converter method for dynamic power allocation based on BES units in a PI-controlled with combined shunt active power filter (SAPF) and BES system have been implemented. In the majority of investigations, the combination of series APF and BES assistance with power systems uses conventional PI control scheme for the VSC [14, 15]. Hence, in this chapter, a combined effect of solar PV, wind energy, and BS integrated series APF is presented and its effects are verified concerning conventional APF. Using the SPV and batteries as storage energy, the overall system was able to achieve bidirectional grid voltage regulation. During voltage disruptions like sag, swell, and interruption, the bidirectional inverter functions as a normal APF, regulating the grid voltage. When the SPV system is using a battery charger, particularly at night time to charge the batteries, it will behave like a conventional switch, and energy will be stored. The proposed technique established the idea of indirect feeder link pathways to regulate voltage regulation, in which as the voltage level rises, a large amount of reverse power flow from WES is automatically regulated by employing the SPV system inverter to charge and store the power.

In summary, the authors highlight the following contributions in this chapter:

- Develop a hybrid energy system using SPV, WES, and BS.
- For voltage sag, swell, and interruption mitigation, a combined hybrid energy system integrated with series APF is proposed.
- Evaluation of the proposed scheme's performance in comparison to traditional APF.

This chapter's final section is organized as follows.

The design of the hybrid energy system model is described in Section 4.2. The modeling of Series APF and its control strategies are covered in Section 4.3. Section 4.4 presents the test scenarios and simulation results. The chapter is briefly concluded in Section 4.5.

4.2 System Topology Description

The recommended structure is shown in Figure 4.1, which also illustrates an energy system that combines with a series APF. The proposed work's detailed structure is depicted in this diagram. A hybrid energy source that is connected across the DC-link side of the series APF at the PCC regulates it. This hybrid energy source regulates the capacitor's DC-link voltage and enhances the APF's filtering capabilities [18–20].

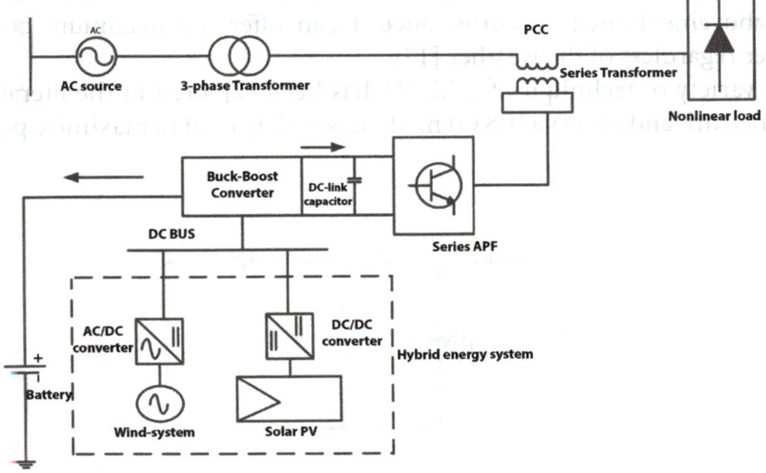

Figure 4.1 Proposed model.

4.2.1 Solar Photovoltaic System

Highly pure solar energy is freely available on the entire planet. Additionally, it produces no greenhouse gas emissions and has a lower carbon footprint. Due to its plentiful supply in the natural world and lack of contamination, solar energy is an excellent replacement for traditional electric power sources. Currently, SPV technologies help to increase system efficiency and lower component costs. The architecture of the SPV technology will be covered in detail in the section that follows.

4.2.1.1 SPV Modeling

A structure derived from previously published research [16] is utilized to simulate the PV array employed in the envisioned system. The PV cell serves as a regulated current source in this technique, as seen in Figure 4.2. The output characteristic of the SPV is derived in Equation 4.1.

$$I_{pv} = N_p \times I_{ph} - N_p \times I_0 \left[\exp\left\{ \frac{q \times V_{pv} + I_{pv} \times R_{se}}{N_s \times AkT} \right\} - 1 \right] \tag{4.1}$$

4.2.1.2 Maximum Power Point Tracking

The fundamental problem with the SPV concept is that the electricity generated is dependent on solar radiation. A maximum power point tracking (MPPT) system with a fast outcome functionality is employed to address the abovementioned concerns since it can offer the maximum output power regardless of the weather [17].

A variety of techniques for MPPT has been explored in the literature. The perturb and observe (P&O) method is widely used to maximize power

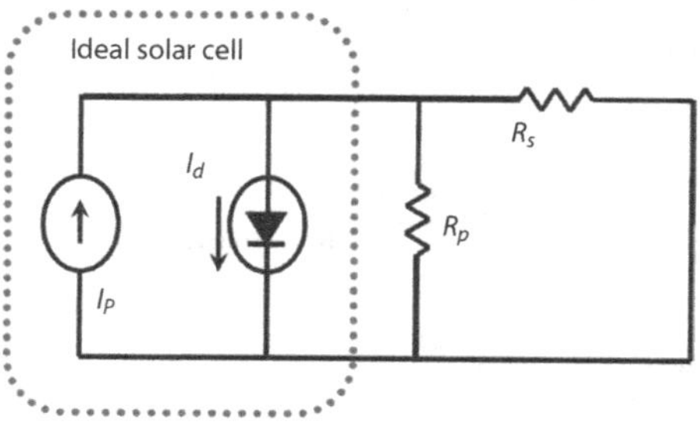

Figure 4.2 SPV equivalent model.

point tracking. Perturb and observe (P&O) provides a simple and quick solution to changes in solar radiation. Because of this, peak energy is produced via P&O regardless of the amount of irradiation. Figure 4.3 displays a flowchart for the P&O MPPT. The SPV system's parameters are listed in Table 4.1 [18].

4.2.1.3 Boost Converter

The boost converter (BC) raises the solar array's output voltage so that it can power a DC-link capacitor. The MPP can also be monitored using the boost converter (BC). Figure 4.4 depicts the layout of the BC. The duty cycle D is used to compute the output DC voltage [19]. Figure 4.5 shows the SPV model with boost converter and MPPT.

4.2.2 Wind Energy System

In the hybrid energy system, WES is another source. The permanent magnet synchronous generators (PMSG)-based WES is used. The wind energy fundamental equation is given in Equation 4.2 [20].

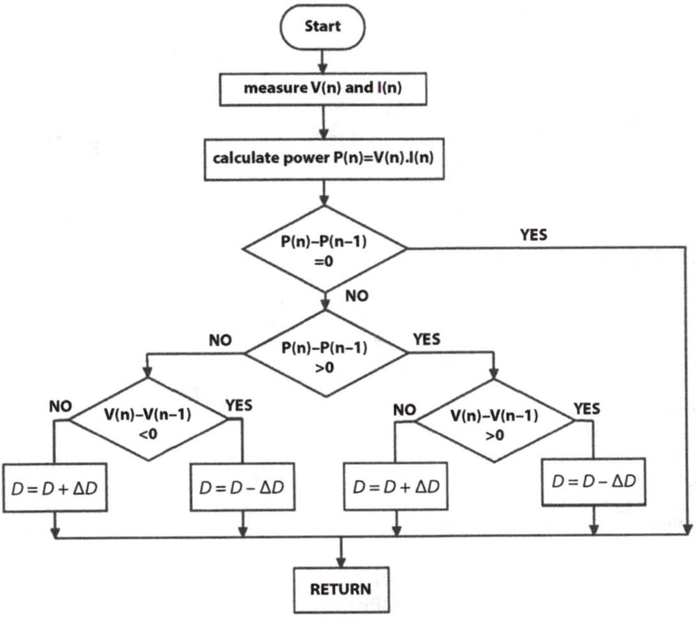

Figure 4.3 Flow chart of P&O MPPT.

Table 4.1 SPV system parameters.

Parameters	Values
Reference temperature (T_r)	25
Diode ideality factor	*A*
Maximum power current (I_{mp})	8.6A
Short circuit current (I_{sc})	10.2A
Open circuit voltage (V_{oc})	90.5V
No. of series-connected cells (N_p)	72
No. of parallel-connected cells (N_s)	01
Voltage output (V_0)	230 V
Supply voltage (V_{pv})	80 V
Maximum power voltage (V_{mp})	81.5V

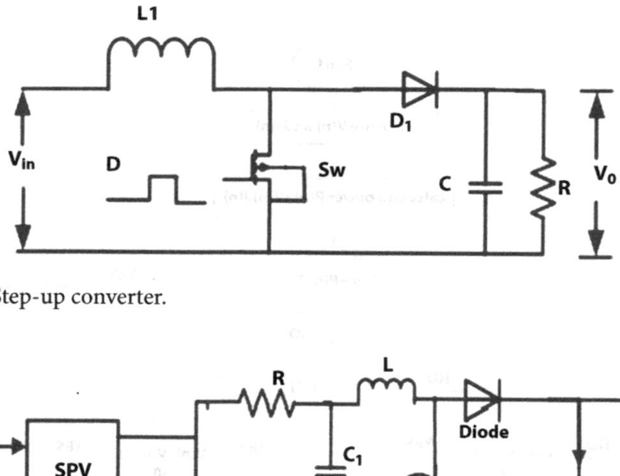

Figure 4.4 Step-up converter.

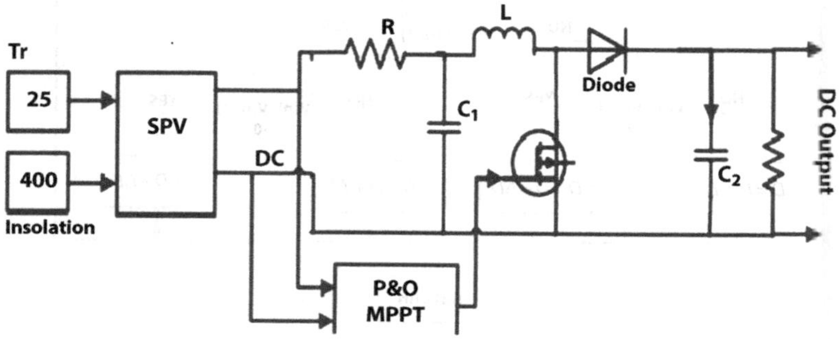

Figure 4.5 SPV model block diagram.

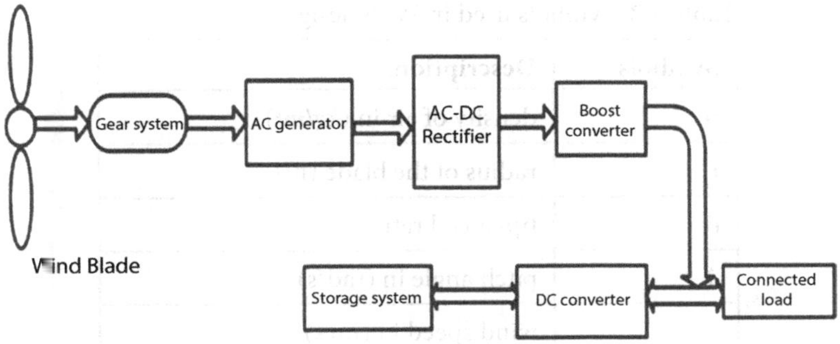

Figure 4.6 WES block diagram.

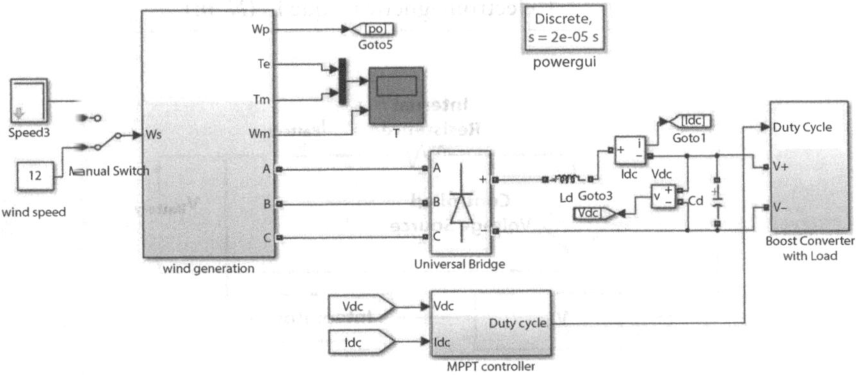

Figure 4.7 Overall WES model.

$$P_0 = \frac{1}{2}\pi\rho C_P(\lambda,\beta)R^2V^3 \tag{4.2}$$

Where P_0 = mechanical power, β = pitch angle, λ = tip-speed ratio, $\lambda = \Omega R/V$, R = blade radius, and ρ = air density. These parameters are used to model and develop the WEC system. Figure 4.6 shows the WES model and Figure 4.7 displays the simulation model of WES. The symbols used in WES are listed in Table 4.2.

4.2.3 Modeling of Battery

In the electric power system, a variety of methods are employed to store energy. Batteries, flywheels, ultra-capacitors, pumped hydro, and the manufacture of hydrogen gas for use in fuel cells or hydrogen internal combustion (IC) engines are among them. Because batteries and flywheels are commercially accessible, they are frequently incorporated at the distribution system level. Lead-acid or

Table 4.2 Symbols used in WES design.

Symbols	Description
ñ	density of air in (kg/m^2)
r	radius of the blade (m)
ë	tip-speed ratio
â	pitch angle in (rad/s)
W_s	wind speed in (r.p.s)
T_m	mechanical torque in (N–m)
T_e	electromagnetic torque in (N–m)

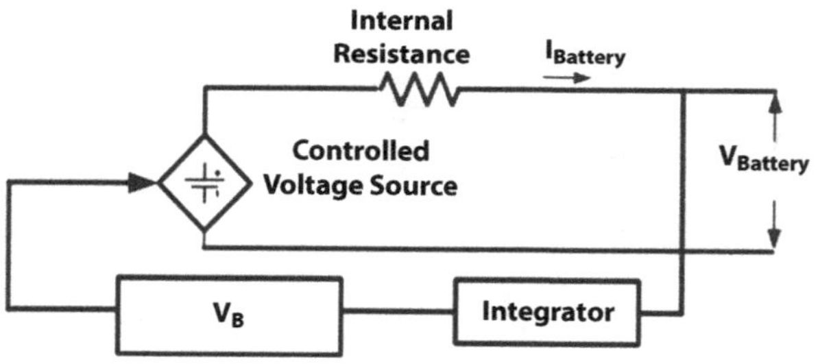

Figure 4.8 Battery model.

flow batteries are commonly used in grid-connected battery banks. Because of their prevalence in everyday electronics, batteries play a crucial function as energy carriers in our modern civilization. Due to the growing popularity of wireless gadgets around the end of the twentieth century, demand for batteries skyrocketed. The battery industry is now classified as a large-scale industry, producing millions of batteries each month. One main development challenge is increasing energy capacity; nevertheless, for consumer items, safety is likely to be considered equally vital today. With the introduction of hybrid electric vehicles to the market, there has been technological advancement in the battery industry, resulting in lower fuel consumption and greenhouse gas emissions. Both business and university research are working on battery development. Figure 4.8 displays an elementary battery architecture.

When the battery is fully depleted and no electricity is flowing, the voltage will be virtually 0. When a current is reintroduced, the voltage drops sharply. This model produces accurate findings and also accurately portrays the battery's behaviour.

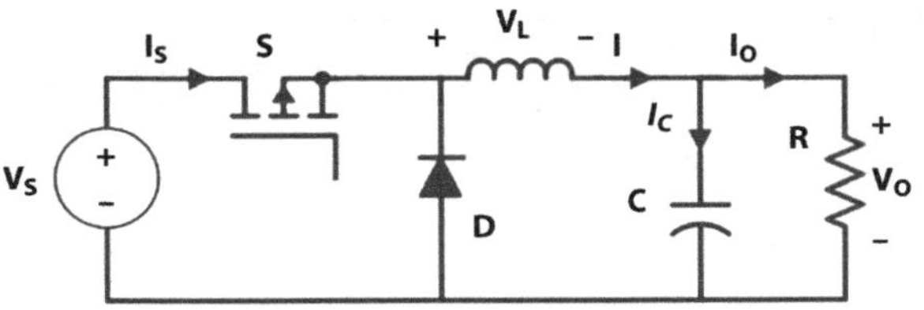

Figure 4.9 Buck converter operation.

4.2.4 Buck-Boost Converter

It is nothing more than a standard DC to DC converter with a predetermined output voltage value. It could be greater or smaller than the input voltage. The buck boost converter is comparable to the fly back system, except instead of using a transformer, it employs a single inductor. There are two different types of converters in a buck-boost converter: buck converters and boost converters. These converters could offer a wider range of output voltages in contrast to input voltages. The fundamental buck-boost converter is depicted in the schematic in Figure 4.9.

Operating Principle

The input inductor, which results in an unanticipated shift in the input current, is what allows the DC to DC converter to work. The inductor absorbs energy from the input and stores it when the switch is switched on. The energy is released when the switch is turned off.

Figure 4.9 shows how a buck converter works. The buck converter's metal oxide silicon field effect transistor (MOSFET) is switched on and the subsequent semiconductor is disabled as a result of the increased square wave frequency. The current passes through the magnetic field, charging C, and powers the load if the magnetic field's current is larger than the first semiconductor's gate terminal's current. The positive voltage applied to the cathode disables the Schottky diode D1.

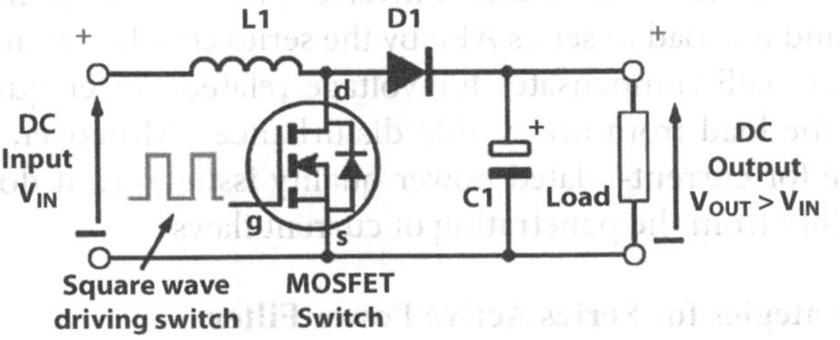

Figure 4.10 Boost converter operation.

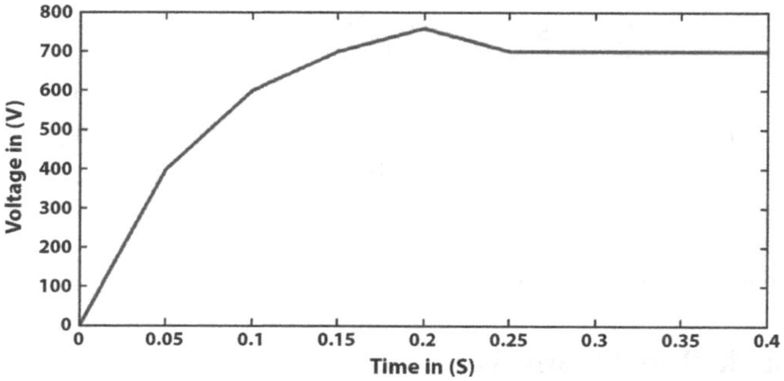

Figure 4.11 Boost converter output voltage.

In the boosting mode shown in Figure 4.10, the primary semiconductor of the aforementioned converter is continuously on while a high-frequency square wave is applied to the second transistor's gate terminal. The other transistor is working while the 'S' MOSFET is turned on and taking input current from the inductor L. The negative terminal charges the magnetic field of the inductor. The D2 diode remains incapable to operate because the anode is on the potential ground because of the robustly conducting MOSFET. Earlier oscillator cycles can be created by charging capacitor C during the ON state, which applies the load to the entire circuit. The output voltage may have a high frequency of ripples when the capacitor C discharges often during the ON period. Figure 4.11 displays the boost converter's output voltage.

4.3 Series Active Power Filter Design

To reduce harmonics, the reactive energy is provided by this DC connection capacitor. The waves in the voltage the inverter produces are eliminated by the inductor and capacitor (LC) filter. As its name suggests, the APF series is coupled in series with the load. Figure 4.12 displays the block diagram of series active filter. The voltage source inverter (VSI) is coupled in series with the source and the load in series APF by the series coupling transformer.

The series APF compensates for voltage-related power quality issues, protecting the load from utility-side disturbances. Moreover, it does not compensate for current-related power quality issues, *i.e.*, it does not prevent the utility from the penetrating of current flaws.

Control Strategies for Series Active Power Filter

The unit vector signals are produced in sequence using the APF Park transformation technique. The measured and real voltage were compared

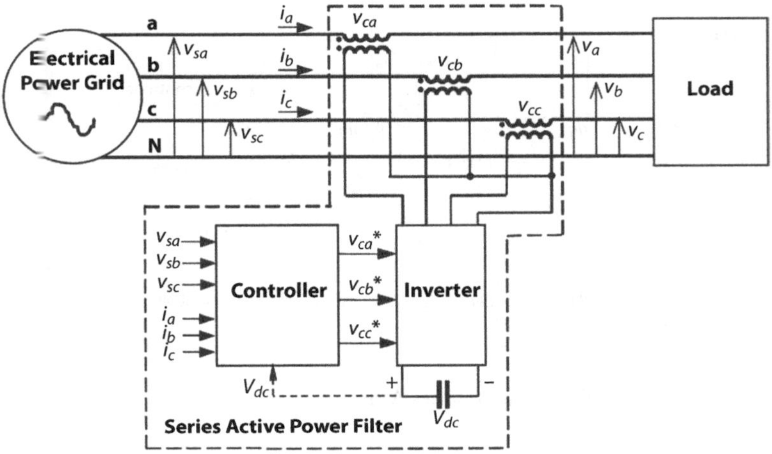

Figure 4.12 Block diagram of series active filter.

in the dq0 reference plane after being translated from abc coordinates to dq0 coordinates. After comparison, both are transformed back into ABC reference frames. The phase locked loop (PLL) required for Park's transformation and inverse Park's transformation can be produced. The series converter's six switches are linked to a pulse width modulation (PWM) generator, which provides synchronized switching pulses. By providing proper switching pulses, the series converter's goals can be met. The generation of the switching pulses is obtained from the three-phase reference signal that is acquired from two controllers. The pulses required for VSI operation are produced by a hysteresis voltage controller by comparing the expected voltage of the output (Vc*) to the real series APF output voltage (Vc). The modeling of series APF control scheme is shown in Figure 4.13.

The series APF detects a voltage drop in a sag condition and injects the necessary voltage through the series coupling transformer. It ensures

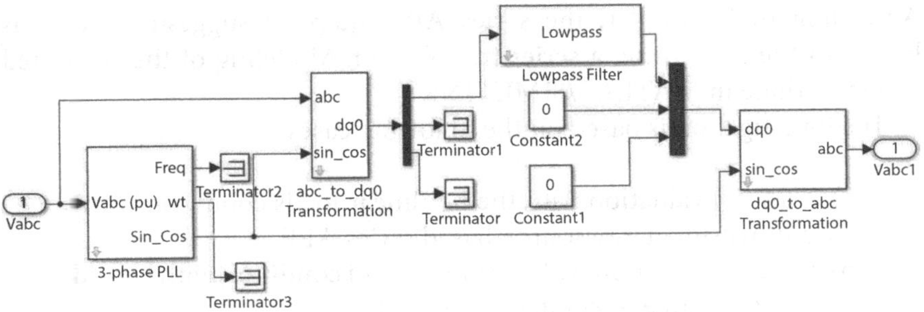

Figure 4.13 Simulation of series APF control scheme.

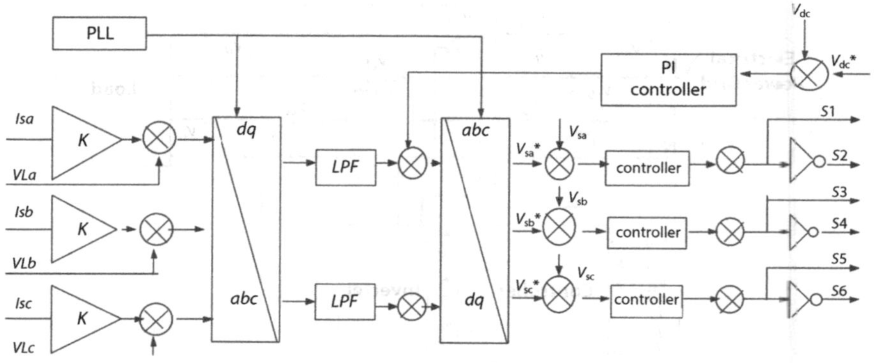

Figure 4.14 Operational diagram of series APF.

that the load terminal is kept at the rated voltage. The series APF recognizes a voltage swell as it happens and injects the compensating voltage to cancel it out, maintaining the evaluated voltage across the load terminal. This method is equivalent to maintaining the rated voltage under uneven source voltage. After the series inverter had injected the compensation voltage in series with the source voltage, it was discovered that the load voltage was similar to the combination of the source and replacement voltages. When the source voltage is non-sinusoidal and unbalanced, the load bus's rated sinusoidal voltage is controlled by injecting a series voltage at the necessary magnitude and phase angle. The system is therefore safeguarded against voltage-related PQ problems including voltage sag, swell, imbalance, flicker, and so forth. Figure 4.14 demonstrates the function of the PI controller.

4.4 Simulation Results

According to Figure 4.1, the series APF-equipped suggested model is linked to the PCC using a series transformer. Modeling of the suggested system is done in MATLAB/SIMULINK^R.

The investigation is based on the following cases.

a) Case 1: evaluation with the nonlinear loads condition without hybrid energy source-based series APF
b) Case 2: evaluation with nonlinear load condition with hybrid energy source integrated series APF

4.4.1 Analysis Under Case 1

In this scenario, the hybrid power sources (HPS) is not connected across the series APF when the nonlinear load is present in order to study the system performance initially.

Sag Compensation

To validate the series APF operation, voltage sag is injected into the system. The series APF is employed at the load terminal to address voltage-related power quality (PQ) abnormalities such voltage sag, swell, inequalities, flicker notches, and so on. The load voltage profile exhibited in Figure 4.15(a) matches the voltage sag observed on the time axis from 0.1 s to 0.3 s. A voltage decrease is detected by the series APF during a sag state, and it injects the necessary voltage through the series coupling transformer. It ensures that the load terminal is kept at the rated voltage. As seen in Figure 4.15(b), when the APF is turned on, it injects compensating voltage into the PCC to make up for load voltage sag. The voltages at the source and the load are therefore equivalent. The voltage of the load following the adjustment is shown in Figure 4.15(c). In order to keep the rated voltage across the load terminal constant, the series APF operates by quickly detecting voltage changes at the source and injecting a compensation voltage. The voltage on the DC-link capacitor is revealed in Figure 4.15(d). The capacitor voltage is practically constant up to 700 V with a settling period of 0.25 s, as shown in the result.

4.4.2 Analysis Under Case 2

The output of a hybrid energy system is combined with a boost converter in this study to achieve the desired voltage. The boost converter is connected to the APF DC-link capacitor to improve performance.

Using a series transformer, the series APF delivers the compensating voltage. The generation-compensating voltage control technique was previously discussed. The sag and swell caused by the load will be mitigated by the compensatory voltage supplied by series APF. Another unique aspect of the system is the employment of a hybrid energy system to maintain a constant DC-link voltage. The proposed technique is ideal for connecting at PCC with various functionalities, allowing for voltage sag correction, load current harmonics suppression and other routine operation.

Sag compensation

To test the effectiveness of the series APF, voltage sag is injected into the system for 0.1 s to 0.3 s of the voltage waveform, shown in Figure 4.16(a).

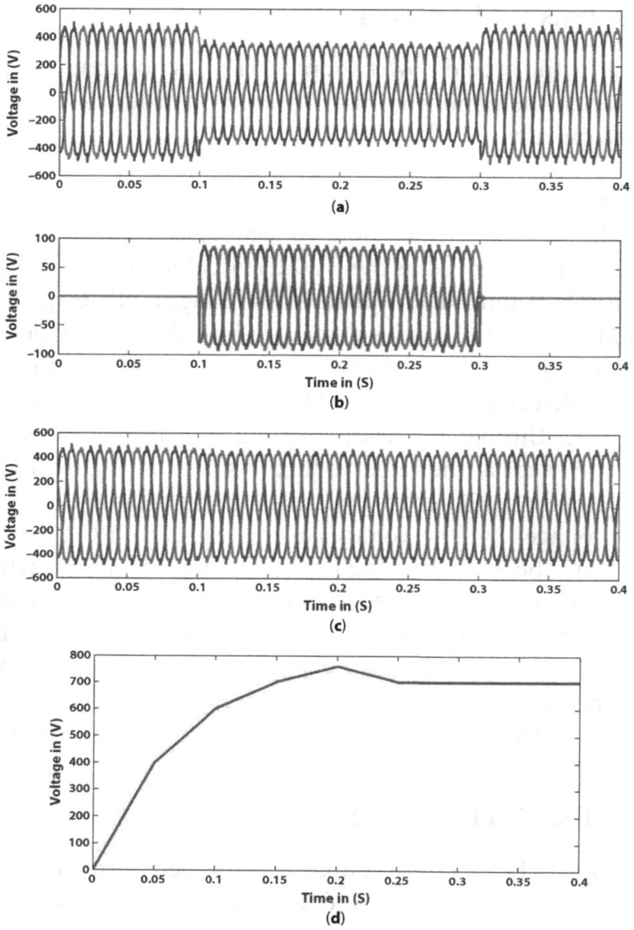

Figure 4.15 Profile obtained Case 1. (a) Load voltage before compensation, (b) compensating voltage injected by APF, (c) load voltage after compensation, and (d) DC-link capacitor voltage.

The required voltage is injected through the series coupling transformer by the series APF, which recognizes the voltage drop occurring during a sag state. It guarantees that the load terminal is kept at the rated voltage. The compensating voltage introduced by the series APF is shown in Figure 4.16(b). The load voltage after compensation is shown in Figure 4.16(c), demonstrating that the sag has been reduced by the injection of compensatory voltage. Figure 4.16(d) displays the voltage across the DC-link

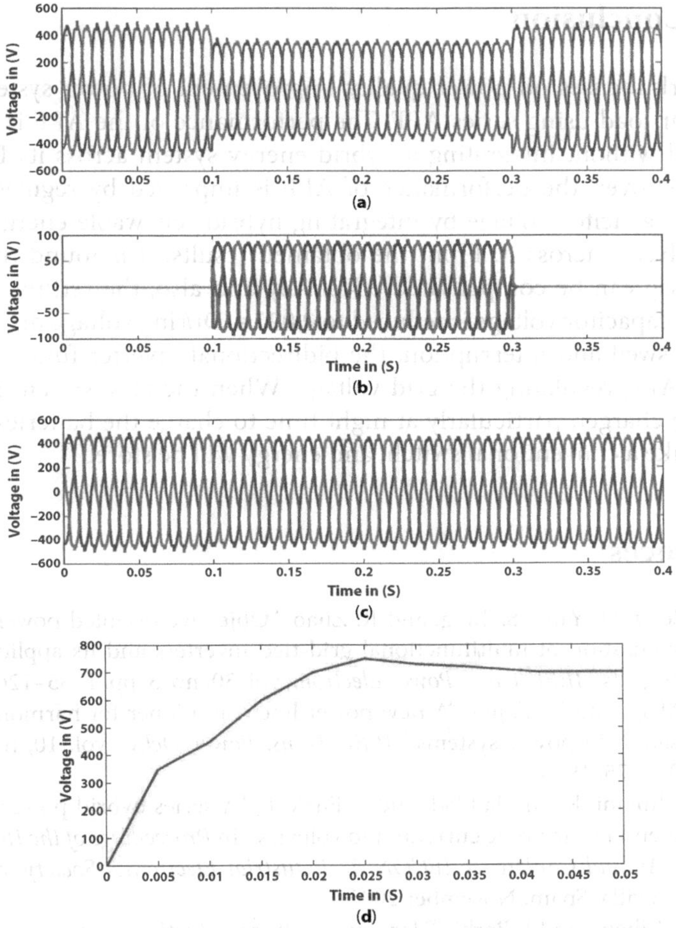

Figure 4.16 Profile obtained Case 2. (a) Load voltage before compensation, (b) compensating voltage injected by APF, (c) load voltage after compensation, and (d) DC-link capacitor voltage.

capacitor. The result demonstrates that the capacitor voltage is nearly constant up to 700 V with a settling time of 0.025 s.

The suggested hybrid energy system with series APF can adjust for current disturbances at the load side while also maintaining a consistent DC-Link capacitor voltage. Nonlinear load is considered when evaluating the suggested APF's performance. The load voltage sag can be efficiently corrected, and the settling time of the DC-link capacitor voltage is reduced to 0.025 s, as shown by the numerical comparison of results.

4.5 Conclusion

This work focuses on power quality improvement of power system with nonlinear load using series APF. The performance of the APF is studied with and without integrating a hybrid energy system across its DC-link side. Moreover, the performance of APF is improved by regulating the DC-link capacitor voltage by integrating hybrid renewable energy-based sources based across it. From the obtained results, it is found that load voltage sag can be compensated effectively and also, the settling time of DC-link capacitor voltage is reduced to 0.025 s. During voltage disruptions like sag, swell and interruption, the bidirectional inverter functions as a normal APF, regulating the grid voltage. When the SPV system is using a battery charger, particularly at night time to charge the batteries, it will behave like a conventional switch, and energy will be stored.

References

1. Z. Zeng, H. Yang, S. Tang, and R. Zhao, "Objective-oriented power quality compensation of multifunctional grid-tied inverters and its application in microgrids," *IEEE Trans. Power Electron.*, vol. 30, no. 3, pp. 1255–1265, 2015.
2. H. Akagi and H. Fujita, "A new power line conditioner for harmonic compensation in power systems," *IEEE Trans. Power Deliv.*, vol. 10, no. 3, pp. 1570–1575, 1995.
3. S. Rahmani, K. Al-Haddad, and F. Fnaiech, "A series hybrid power filter to compensate harmonic currents and voltages," in *Proceedings of the IEEE 2002 28th Annual Conference IECON 02 Industrial Electronics Society*, pp. 644–649, Sevilla, Spain, November 2002
4. E. R. Ribeiro and I. Barbi, "Harmonic voltage reduction using a series active filter under diffferent load conditions" *IEEE Trans. Power Electron.*, vol. 21, no. 5, pp. 1394–1402, 2006.
5. F. Z. Peng, H. Akagi, and A. Nabae, "A new approach to harmonic compensation in power systems-a combined system of shunt passive and series active fifilters," *IEEE Trans. Ind. Appl.*, vol. 26, no. 6, pp. 983–990, 1990.
6. S. Janpong, K. L. Areerak, and K. N. Areerak, "A literaturesurvey of neural network applications for shunt active power filters," *Int. J. Electr. Comput. Eng.*, vol. 5, no. 12, 2011.
7. B. Singh, G. Bhuvaneswari, and S. R. Arya, "Review on powerquality solution technology," *Asian Power Electron. J.*, vol. 6, no. 2, 2012.
8. C. Kumar and M. K. Mishra, "An improved hybrid DSTATCOM topology to compensate reactive and nonlinear loads," *IEEE Trans. Ind. Electron.*, vol. 61, no. 12, pp. 6517–6527, 2014

9. S. M. Mao, Q. Chen, R. Li, and X. Cai, "Control of a cascaded STATCOM with battery energy storage system under unbalanced and distorted grid voltage conditions," *J. Renew. Sustain. Energy*, vol. 9, no. 4, pp. 044104, Jul. 2017.
10. T. Renuka and G. Kesavarao, "STATCOM with battery and super capacitor hybrid energy storage system for enhancement of voltagestability," *Indones. J. Electr. Eng. Comput. Sci.*, vol. 5, no. 2, pp 250–259, Feb. 2017.
11. N. N. Li, Y. Q. Liu, J. Z. Wang, Y. C. Ji, and B. Xie, "Dynamic allocation method of DC side power based on the SoC of battery for STATCOM/BESS," *Electr. Power Syst. Res.*, vol. 125, pp. 141–149, Aug. 2015.
12. A. Chakraborty, S. K. Musunuri, A. K. Srivastava, and A. K. Kondabathini, "Integrating STATCOM and battery energy storage system for power system transient stability: a review and application," *Adv. Power Electron.*, vol. 2012, pp. 676010, Dec. 2012.
13. C. Enang and B. K. Johnson, "Enhanced modular multilevel converter based STATCOM with hybrid energy storage," in *Proceedings of 2019 IEEE Power & Energy Society General Meeting (PESGM)*, Atlanta, GA, USA, 2019, pp. 1–5
14. S. Pati, S. K. Kar, K. B. Mohanty, and S. Mishra, "A sliding mode controller based STATCOM with battery storage for voltage and frequency stabilization in a micro-grid," in *Proceedings of 2016 International Conference on Circuit, Power and Computing Technologies (ICCPCT)*, Nagercoil, 2016, pp. 1–7.
15. Pathak, P. K., Yadav, A. K., & Alvi, P. A. (2020). Advanced solar MPPT techniques under uniform and non-uniform irradiance: a comprehensive review. *J. Sol. Energy Eng.*, 142(4).
16. Mao, M., Cui, L., Zhang, Q., Guo, K., Zhou, L., & Huang, H. (2020). Classification and summarization of solar photovoltaic MPPT techniques: A review based on traditional and intelligent control strategies. *Energy Rep.*, 6, 1312-1327.
17. Samal, S., & Hota, P. K. (2017). Power quality improvement by solar photovoltaic/wind energy integrated system using unified power quality conditioner. International *Int. J. Power Electron. Drive Syst.*, 8(3), 1424.
18. Barik, P. K., Shankar, G., &Sahoo, P. K. (2020). Power quality assessment of microgrid using fuzzy controller aided modified SRF based designed SAPF. *Int. Trans. Electr. Energy Syst.*, 30(4), e12289.
19. S. Samal, A. Hota, P.K. Hota, P.K. Barik, Harmonics and voltage sag compensation of a solar pv-based distributed generation using MSRF-based UPQC, in *Innovation in Electrical Power Engineering, Communication, and Computing Technology. Lecture Notes in Electrical Engineering*, 630 (Springer, Berlin, 2020).
20. S. Samal, P.K. Hota, P.K. Barik, Power quality assessment of a solar PV and fuel cell based distributed generation system using unified power quality conditioner. *Int. J. Ambient Energy* 1–34 (2020).

5

Dynamic Modeling of Drone Control with MATLAB Simulation

Suman Lata Tripathi

School of Electronics and Electrical Engineering, Phagwara, Punjab, India

Abstract

This chapter focuses on dynamic modeling of drone using MATLAB. The control of diverse drone movements is vital for understanding and managing drone flight dynamics. The main objective is to replicate movements like ascending, descending, forward-backward, and lateral translations through a comprehensive simulation model within MATLAB. This simulation will accurately represent the drone's flight dynamics and interactions within its environment. The understanding of drone maneuverability and control is serving as a foundational step towards developing more sophisticated control systems for these aerial platforms by facilitating advancements in drone technology across various industries.

Keywords: Flight control, drone control, simulation in MATLAB, flight dynamics, six degrees of freedom, dynamic systems modelling, parameter analysis, aerospace technology

5.1 Introduction

Drones, also known as unmanned aerial vehicles (UAVs), have become increasingly popular in recent years due to their versatility and usefulness in a wide range of industries (Shelare *et al.*, 2021) [1]. These aerial vehicles are equipped with advanced technology and can be used for a variety of purposes, including surveillance, reconnaissance, agriculture, disaster management, and filmmaking. Drones are particularly valuable because

Email: tri.suman78@gmail.com

Tamal Roy, Suman Lata Tripathi and Souvik Ganguli (eds.) Linear and Nonlinear System Modeling, (67–88) © 2024 Scrivener Publishing LLC

they can access remote or dangerous areas, collect real-time data, and perform tasks with precision and efficiency.

Understanding the complex dynamics of drone flight is crucial for optimizing their performance and ensuring safe and reliable operation. Drones can move in six directions (six degrees of freedom or 6-DOF), which means they can travel up, down, forward, back, left, and right, and they can also rotate in all three dimensions (roll, pitch, and yaw) (Najm and Ibraheem, 2019) [1]. Comprehending these 6-DOF is essential for effective drone control because it directly affects how we can manage the drone's position, orientation, and overall flight dynamics. This understanding is the foundation for developing precise control strategies, enhancing operational capabilities, and navigating the challenges inherent in drone flight.

MATLAB's Simulink is a powerful platform for modeling, simulating, and analyzing dynamic systems, making it an ideal tool for understanding the complexities of drone flight. This chapter aims to provide a comprehensive understanding of the parameters governing drone flight, enabling proficient navigation of the MATLAB and Simulink environment.

5.2 Tool Description

Tools play an integral role across various industries, enabling efficiency and precision in tasks. These instruments range from simple handheld devices to complex machinery, designed to assist in construction, manufacturing, repair, or creative endeavors. Each tool serves a specific function, aiding individuals in accomplishing tasks with greater accuracy and speed. Their innovation and evolution have significantly contributed to advancing human capabilities, allowing for enhanced productivity and the realization of drone model. The following tools are required for designing of drone model:

MATLAB is a powerful numerical computing environment widely used in engineering, mathematics, and scientific research. It provides a comprehensive set of tools for data analysis, visualization, and algorithm development. With its user-friendly interface and extensive libraries, MATLAB allows users to solve complex mathematical problems, simulate systems, and create intricate models with ease. Its versatility in handling matrices, functions, and algorithms makes it an indispensable tool for professionals and academics alike.

Simulink is an extension of MATLAB, specifically designed for modeling, simulating, and analyzing dynamic systems. It offers a graphical environment for designing and simulating multidomain systems using block diagrams. Engineers and researchers use Simulink for various applications such as control systems, signal processing, communications, and more. Its intuitive interface allows users to model complex systems visually by connecting blocks representing components or functions. Simulink's simulation capabilities enable real-time analysis and testing of dynamic systems before implementation, making it an invaluable tool for system design and validation.

The *Simulink support package for Parrot minidrone* serves as a vital link between Simulink, an advanced simulation and modeling environment, and Parrot's minidrone platform. This package provides users with an extensive toolkit and various blocks within Simulink, enabling seamless communication and control of the Parrot minidrone. By merging Simulink's intuitive visual interface with the minidrone's capabilities, this package empowers engineers, researchers, and enthusiasts to create, simulate, and implement intricate control algorithms for the drone. Users can refine and experiment with control strategies virtually before deploying them onto the physical minidrone, expediting development and encouraging innovation in drone control applications.

5.3 Methodology

In the process of designing a drone control system using MATLAB, the understanding of the 6-DOF in drone flight (thrust or up/down, pitch or forward/back, roll or left/right, and yaw or twist left and right) is essential for applying effective engineering. This knowledge serves as the foundation for developing a control system that enables the drone to hover and move seamlessly.

The design of a control system for a drone involves a thorough grasp of hardware components and sensors, such as ultrasound, camera, pressure sensor, and inertial measurement unit (IMU). Addressing the underactuated nature of the system, with four actuators for 6-DOF, presents a challenge. The configuration and spin directions of motors are pivotal, requiring a motor mixing algorithm to convert intuitive commands into precise motor speeds. The final control system, built and simulated in Simulink, undergoes tuning and testing before generating flight code for the onboard microcontroller. This methodology leverages sensor data for state estimation and deploys algorithms for precise motor control, ensuring desired drone movements in three-dimensional (3D) space.

5.4 Overview of the Drone Control System

To truly understand the flow of the design of the control system, the image above provides an overview of the control problem at hand. The system involves a drone as the plant and four actuators responsible for introducing forces and torques into the system. The fundamental question revolves around determining the optimal inputs to achieve the desired output—specifically, manipulating the four motors in precise ways to enable the drone to rotate and maneuver in 3D space. Supporting this endeavor are a set of sensors employed to directly or indirectly estimate the minidrone's state, encompassing angular position and rates, altitude, and horizontal velocity. The states subject to estimation are contingent on the chosen control architecture and the objectives pursued, details of which are elucidated in Figure 5.1 and in subsequent discussions.

Having the required knowledge of the system's state and a clear understanding of the desired minidrone behavior, a controller is developed. This controller, essentially an algorithm executed in software, takes the set point and estimated state as inputs and calculates precise motor commands to inject the requisite forces and torques into the system.

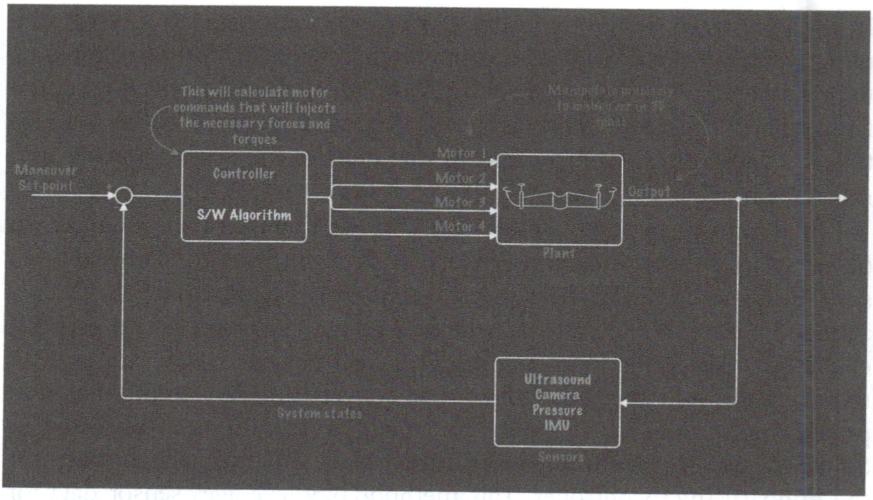

Figure 5.1 Block diagram of the drone control system.

5.5 Overview of the Drone Control System in MATLAB Simulink

The drone control system in MATLAB Simulink [3], comprises four primary block groups. Firstly, the flight command is represented, housing the signal builder command block to regulate flight directives. Secondly, the drone flight control system is depicted, serving as a central component for managing the drone's movements. The third block group encompasses the simulation model, featuring three blocks: nonlinear airframe, environment, and sensors, collectively contributing to a comprehensive simulated environment. Lastly, the flight visualization block provides a visual representation of the drone's flight dynamics. This organized model structure facilitates a systematic and efficient approach to drone control and simulation within the MATLAB Simulink framework.

5.5.1 Flight Command

The flight command block group serves as a crucial component in the drone's control system architecture, encapsulating the signal builder block that intricately manages the command controls for the 6-DOF parameters: thrust, roll, yaw, and pitch. Within the signal builder block, precise instructions are crafted to dictate the drone's movements in 3D space, allowing for seamless control over its translational and rotational dynamics. Notably, this block goes beyond basic commands by also defining orientation and position references, ensuring a comprehensive and sophisticated approach to directing the drone's flight path. The integration of the signal builder within the flight command block group exemplifies a meticulous design strategy for orchestrating the drone's precise and dynamic maneuvers in response to user input or autonomous navigation requirements.

5.5.2 Flight Control System

The flight control system block serves as a pivotal component in directing the drone's flight maneuvers. It receives commands from the flight command, specifying the desired movements of the drone. Simultaneously, it collects real-time feedback from sensors and processes image data to adjust the drone's actions based on its actual position and environmental conditions. Acting as a vigilant overseer, this block includes a fail-safe flag designed to halt the actuators in the event of software-related issues. The output signals from this block are essential instructions sent to the actuators, precisely governing the motor actions. Meanwhile, the flag acts as a safety measure, ensuring the drone's flight remains stable and secure, particularly in unforeseen circumstances.

5.5.3 Simulation Model

Simulation models serve as an invaluable tool for understanding, analyzing, and replicating the behavior, functionality, and performance of a real-life drone. It involves intricate algorithms, physics-based calculations, and sensory inputs to imitate flight dynamics, navigation, control systems, and environmental interactions. These models enable testing and refining drone designs, predicting performance under various conditions, and developing efficient control strategies, fostering innovation and advancements in drone technology across industries like aerial photography, logistics, surveillance, and more.

5.5.4 Flight Visualization

The flight visualization block group is an integral component of the drone control system, providing real-time feedback on the drone's state and enabling operators to effectively monitor its performance and address any potential issues. This block group seamlessly integrates with the drone's sensors and actuators to gather comprehensive data on its position, orientation, altitude, and other critical parameters. The acquired data is then processed and visualized using a variety of graphical elements, including gauges, charts, and maps, providing a clear and intuitive representation of the drone's status. The flight visualization block group offers several advantages to drone operators, including enhanced situational awareness, improved performance monitoring, and facilitated troubleshooting. By providing a comprehensive overview of the drone's state, operators can swiftly identify and address any emerging issues, ensuring the drone's safe and reliable operation.

5.5.5 Result and Discussion

This section reveals the results and discussion on the performance of the drone control system and simulation, focusing on the key parameters of thrust, pitch, roll, and yaw. Through simulations and tabulated data have systematically explored the influence of these critical parameters on the drone's navigation capabilities. By examining the dynamic interplay between thrust and the drone's vertical movement, as well as the impact of pitch, roll, and yaw on its orientation, we aim to gain insights into the effectiveness and responsiveness of our drone control system.

5.5.6 Varying the Values of Thrust Parameter of the Drone Flight Control

Thrust is a critical parameter for drones, enabling them to move freely in 6DOF. This means the drone can not only fly forward and backward but also move up and down, left and right, and even rotate on its axis. Without sufficient thrust, a drone would be unable to achieve lift-off, let alone maneuver with agility. Figure 5.2 presents a simulated result of drone control model. Table 5.1 depicts the drone flight position with varying the values of thrust parameter.

In the provided simulation results, the variations in the thrust parameter of the drone flight control system have a clear and direct impact on the drone's vertical movement. When the thrust parameter is set to 0, the drone maintains its position, indicating a balanced state with no net upward or downward force. Subsequently, when the thrust parameter is set to –2, a negative value, the drone exhibits an upward movement, suggesting that the propulsion system is generating more thrust than needed to counteract gravity. This behavior aligns with the intuitive expectation that a higher thrust value would cause the drone to ascend.

Conversely, as the thrust parameter transitions from –2 to –0.3, a negative but less extreme value, the drone begins to descend. This downward

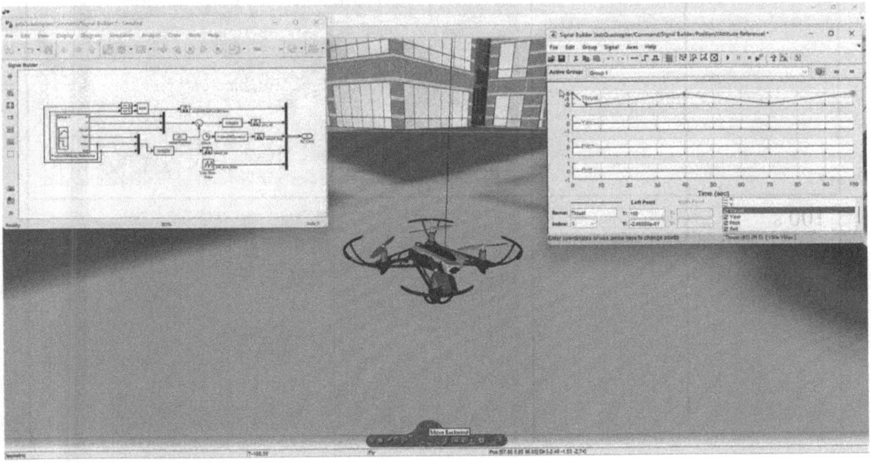

Figure 5.2 Simulation results of the drone control model.

Table 5.1 Varying the values of thrust parameter of the drone flight control.

Time	Values	Simulation
0 s	0	
5 s	–2	
40 s	–0.3	
70 s	–2	
100 s	0	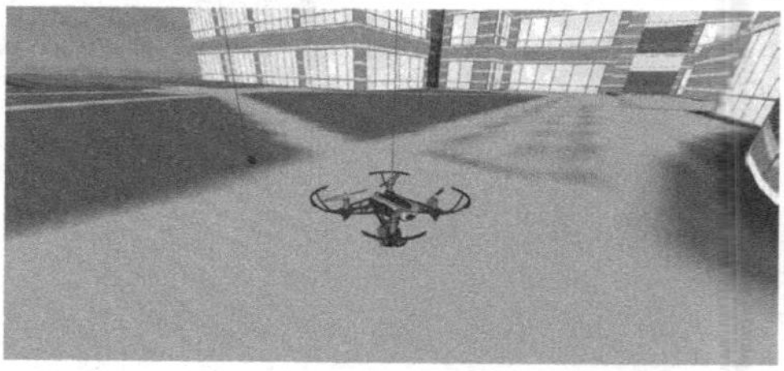

movement implies that the thrust generated by the propulsion system is no longer sufficient to counteract the force of gravity, resulting in a net downward force. The observed behavior is consistent with the fact that a decrease in thrust leads to a descent. Finally, as the thrust parameter returns to 0, the drone resettles to its initial position. This cyclic behavior illustrates the responsiveness of the drone to changes in the thrust parameter, serving as a crucial insight for drone flight control adjustments and emphasizing the direct correlation between thrust values and the drone's vertical movement.

5.5.7 Varying the Values of Pitch Parameter of the Drone Flight Control

The pitch parameter plays an important role in a drone's ability to achieve complete movement within the 6-DOF. As one of the fundamental parameters in drone flight control, the pitch parameter specifically governs lateral motion along the horizontal axis. By manipulating the values for pitch parameters, the drone motors will synchronously operate to seamlessly traverse backward and forward positions, enabling a versatile and dynamic range of motion essential for various applications. Figure 5.3 depicts the MATLAB-based drone model for pitch parameter.

Table 5.2 shows the variation of the values of pitch parameters of the drone flight control that will exhibit the functionality of the designed control system in line with the pitch parameter. The simulations are also shown

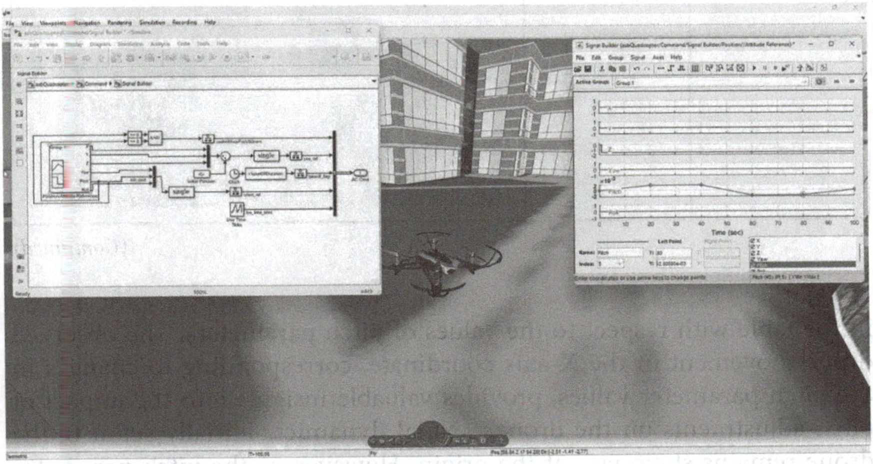

Figure 5.3 Drone model for pitch parameter.

Table 5.2 Varying the values of pitch parameter of the drone flight control.

Time	Values	Simulation
0 s	0	
20 s	0.002	
40 s	0.002	
60 s	-0.002	

(Continued)

in the table with respect to the values of pitch parameters. The observed drone movement in the X-axis coordinate, corresponding to changes in the pitch parameter values, provides valuable insights into the impact of pitch adjustments on the drone's spatial dynamics. Initially set at 0, the drone remains stationary at the origin. However, as the pitch parameter

Table 5.2 Varying the values of pitch parameter of the drone flight control. (*Continued*)

Time	Values	Simulation
80 s	-0.002	
100 s	0	

decreases to −0.002, the drone exhibits a backward movement, indicating a pitch adjustment that influences the drone to tilt backward. This backward tilt results in a displacement in the negative X-axis direction, effectively moving away from the starting point.

Conversely, when the pitch parameter transitions from −0.002 to 0.002, the drone responds by tilting forward. This adjustment prompts the drone to move in the positive X-axis direction, indicating a forward motion. The directionality of the drone's movement aligns with the sign of the pitch parameter, demonstrating a clear correlation between pitch adjustments and the drone's spatial behavior. The return to a pitch parameter value of 0 corresponds to the drone returning to its original position, signifying a reset or neutralization of the pitch influence and, consequently, the cessation of any forward or backward movement along the X-axis.

5.5.8 Varying the Values of Roll Parameter of the Drone Flight Control

Another essential parameter for a drone's 6-DOF movement is roll. This angular movement allows the drone to tilt to its left and right sides. This tilting ability is crucial for maneuvering the drone, enabling it to turn, bank, and avoid obstacles. Without roll control, the drone would fly in a straight line, severely limiting its ability to navigate and perform complex maneuvers. Figure 5.4 shows the drone flight control model for roll parameter.

The observed variations in the roll parameter values of the drone's flight control, as outlined in Table 5.3, indicate significant effects on the drone's lateral movement along the Y-axis coordinate. Initially, when the roll parameter is set to 0, the drone remains stationary at the origin. However, as the roll parameter increases to 0.002, a positive value, the drone exhibits a consistent lateral movement to the right over the course of 40 s. This behavior is in line with the fundamental principle of drone flight control, where adjustments in the roll parameter contribute to the drone's angular rotation around its longitudinal axis, consequently influencing its lateral displacement.

Conversely, when the roll parameter transitions from 0.002 to –0.002, a negative value, the drone's motion reverses, leading to a lateral movement to the left. This behavior aligns with the expectation that a negative roll parameter induces a rotation in the opposite direction, effectively steering the drone to the left along the Y-axis. The subsequent return to a roll parameter value of 0 results in the drone retracing its path back to the starting point. These observed movements underscore the critical role of the roll parameter in controlling the drone's lateral displacement, showcasing its direct influence on the direction and magnitude of the drone's movement in the Y-axis coordinate.

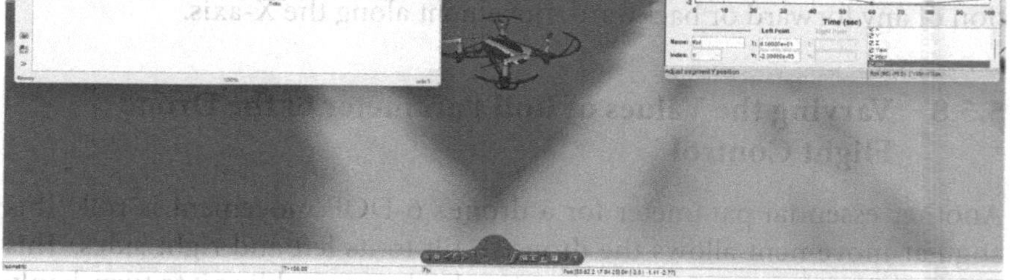

Figure 5.4 Drone flight model for roll parameter.

Table 5.3 Varying the values of roll parameter of the drone flight control.

Time	Values	Simulation
0 s	0	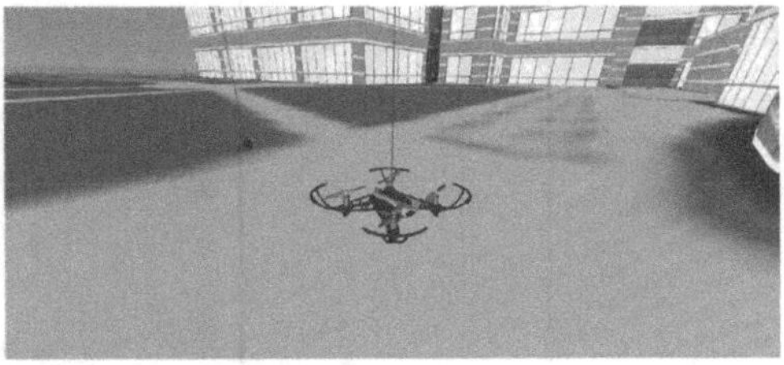
20 s	0.002	
40 s	0.002	
60 s	–0.002	
80 s	–0.002	

(Continued)

Table 5.3 Varying the values of roll parameter of the drone flight control. (*Continued*)

Time	Values	Simulation
100 s	0	

5.5.9 Varying the Values of Yaw Parameter of the Drone Flight Control

Within the 6-DOF that define a drone's movement, yaw plays a critical role. This parameter controls the drone's rotational movement around its vertical axis, allowing it to twist left and right. This ability to twist and turn is vital for precise maneuvering, enabling the drone to change direction quickly and accurately. Without yaw control, the drone would only be able to fly forward and backward, severely limiting its ability to navigate complex environments and perform tasks like aerial photography and videography. Figure 5.5 shows a drone flight control model for yaw parameter.

The provided Table 5.4 illustrates the impact of varying the yaw parameter values in the drone flight control system at different time intervals. Yaw,

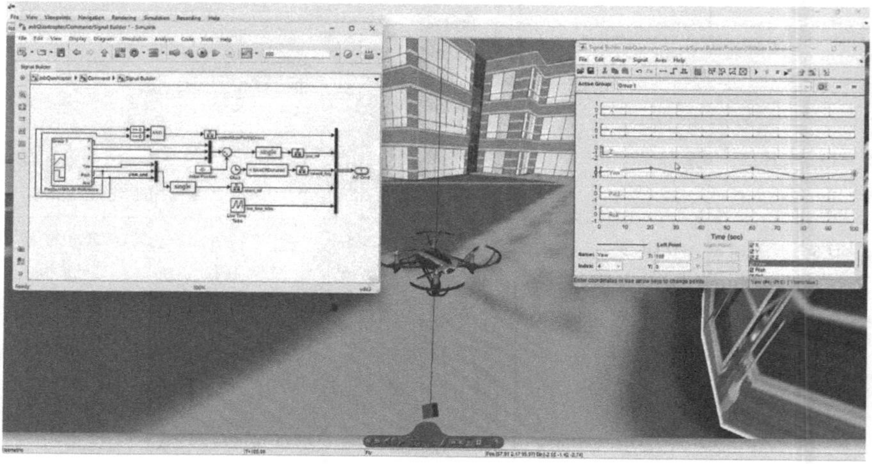

Figure 5.5 Drone flight control model for yaw parameter.

Table 5.4 Varying the values of yaw parameter of the drone flight control.

Time	Values	Simulation
0 s	0	
20 s	1	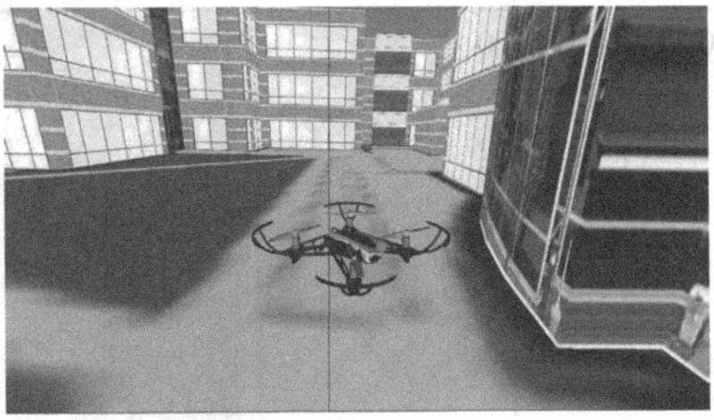
40 s	−1	
60 s	1	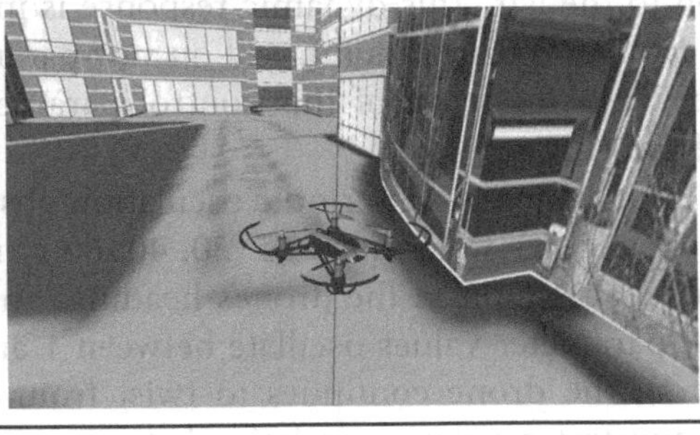

(Continued)

Table 5.4 Varying the values of yaw parameter of the drone flight control. (*Continued*)

Time	Values	Simulation
80 s	-1	
100 s	0	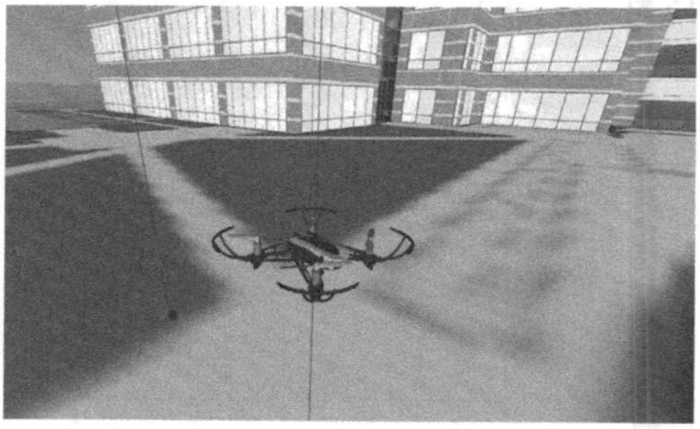

representing the rotation around the vertical axis, plays a crucial role in determining the drone's orientation and heading. As observed in the simulation, when the yaw parameter is set to 0, the drone maintains its initial heading. When the yaw value transitions from 0 to 1, the drone exhibits a clockwise twist or rotation to the right. Conversely, when the yaw value changes from 1 to −1, the drone initiates an anti-clockwise twist or rotation to the left. This dynamic response is indicative of the drone's ability to change its orientation based on yaw parameter variations, highlighting the direct influence of this control parameter on the drone's rotational behavior.

Furthermore, the periodic oscillation observed in the yaw values (0, 1, −1) at regular time intervals (20, 40, 60 s) reflects a deliberate and controlled alteration in the drone's heading. This pattern repeats, and as the yaw parameter values oscillate between 1 and −1 at subsequent intervals (80 s), the drone continues to twist from right to left. The simulation demonstrates a clear correlation between yaw parameter adjustments and the resulting rotational behavior of the drone. It emphasizes the practical significance of fine-tuning yaw parameters in drone flight control to achieve specific orientation changes, allowing for precise maneuvering and navigation in different directions.

5.5.10 Varying the Values of Thrust, Pitch, Roll, and Yaw Parameter of the Drone Flight Control

A drone's ability to move freely in all directions comes down to four key parameters: thrust, pitch, yaw, and roll. Thrust controls the drone's altitude, with more thrust sending it higher and less pulling it down. Pitch and yaw alter its direction, tilting the body forward and backward for pitch, and rotating left and right for yaw. Finally, roll tilts the drone sideways, allowing it to bank and turn. Mastering these parameters allows the drone to move with incredible precision and agility, making it capable of traversing complex environments and performing intricate maneuvers.

The simulation results, presented in Table 5.5 and 5.6, illustrate the effectiveness of a synchronized control system governing thrust, roll, pitch, and yaw parameters for a drone. By modulating these parameters at different time intervals, the drone exhibits controlled movement in all 6-DOF. The coordinated adjustments in thrust, roll, pitch, and yaw enable the drone to navigate through changes in altitude, orientation, and rotational movements. The tables provide a clear visualization of the drone's dynamic response, showcasing the practical application and capability of the designed control system for comprehensive and precise drone navigation.

Table 5.5 Varying the values of all parameters of the drone flight control.

Time	Thrust value	Pitch value	Roll value	Yaw value
0 s	0	0	0	0
20 s	–1	0.002	0.002	0.5
40 s	–0.8	0.002	0.002	–0.5
60 s	–1.2	–0.002	–0.002	0.5
80 s	–1	–0.002	–0.002	–0.5
100 s	0.2	0	0	0

Table 5.6 Varying the values of all parameters of the drone flight control.

Time	Simulation
0 s	
20 s	
40 s	
60 s	

(*Continued*)

Table 5.6 Varying the values of all parameters of the drone flight control. (*Continued*)

Time	Simulation
80 s	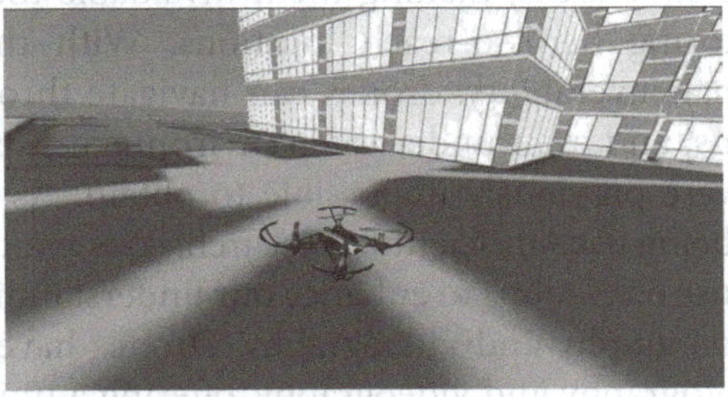
100 s	

5.6 Applications

Drones, also known as UAVs, have risen from futuristic fantasy to commonplace reality. These versatile machines, equipped with rotors and remote control, are buzzing into action across diverse fields, transforming industries and redefining possibilities. Surveillance, environmental monitoring and photography and videography industries are some applications in drone control [3, 4].

Surveillance. Drones have become the eyes in the sky, transforming surveillance and monitoring. These UAVs equipped with powerful cameras and sensors can gather real-time data from inaccessible or dangerous areas, providing a comprehensive view that traditional methods struggle to match. Imagine soaring above a crowd, capturing faces and activities with unparalleled detail, or swiftly surveying vast landscapes, pinpointing anomalies hidden from ground-based observers. This is the power of drone surveillance, offering unparalleled awareness for security, disaster response, environmental protection, and even wildlife conservation. From patrolling borders to tracking endangered species, drones are revolutionizing the way we monitor our world, one silent flight at a time.

Environmental monitoring. Drones have the ability to access remote and inaccessible areas, making them invaluable tools for gathering data and monitoring hard-to-reach locations. With their advanced technology and maneuverability, drones can navigate through dense forests, rugged terrains, and even underwater environments, providing valuable insights that were previously impossible to obtain. As a result, they are increasingly being used in scientific research, conservation efforts, and emergency response operations to enhance our understanding of the world around us.

Photography and videography. Drones have revolutionized the world of photography and videography, offering a unique perspective and capabilities that were previously unimaginable. These flying machines are no longer just for hobbyists, they have become essential tools for professionals and aspiring creators alike. Drones can capture stunning aerial photos and videos that would be impossible to get with traditional methods. With their ability to maneuver in tight spaces and capture shots from various angles, drones have opened up new creative possibilities for photographers and filmmakers. Additionally, their compact size and portability make them convenient for on-the-go shooting, allowing professionals to easily capture breathtaking visuals in remote or challenging locations.

As drone technology continues to evolve, its applications are bound to expand even further. With advancements in drone technology, we can expect to see improved safety features such as collision avoidance systems and enhanced battery life, enabling longer flight times and more efficient operations. Furthermore, the integration of artificial intelligence [5] and machine learning algorithms into drones could revolutionize industries like agriculture, where they can be used for precision farming and crop monitoring [6]. As the potential of drones continues to be explored, it is clear that they will play a significant role in shaping our future.

5.7 Conclusion

In conclusion, this chapter provided an overview of the design of drone control and simulation using MATLAB and its Simulink environment. The chapter provided a comprehensive understanding of the parameters governing drone flight, enabling proficient navigation of the MATLAB and Simulink environment. An effective control system capable of precisely maneuvering the drone was designed and implemented. Through the utilization of Simulink, realistic drone simulations were conducted to evaluate control strategies and optimize performance under various conditions. The chapter also presented an opportunity for the readers to overcome the challenge of unfamiliarity with the platform required to operate a drone by thoroughly mastering the MATLAB and Simulink environment, enabling them to effectively control and simulate drone flight.

References

1. Shelare SD, Aglawe KR, Waghmare SN, Belkhode P (2021b) Advances in water sample collections with a drone – A review. *Mater. Today Proc.* 47:4490–4494. https://doi.org/10.1016/j.matpr.2021.05.327
2. Najm AA, Ibraheem IK (2019) Nonlinear PID controller design for a 6-DOF UAV quadrotor system. *Eng. Sci. Technol.*, an International Journal 22(4):1087–1097. https://doi.org/10.1016/j.jestch.2019.02.005
3. MATLAB: https://in.mathworks.com/discovery/drone-simulation.html
4. Goyal, N., Sharma, S., Rana, A. K., & Tripathi, S. L. (Eds.). (2022). *Internet of Things: Robotic and Drone Technology*. CRC Press, Boca Raton, UK.
5. Thillaiarasu, N., Tripathi, S. L., & Dhinakaran, V. (Eds.). (2022). *Artificial Intelligence for Internet of Things: Design Principle, Modernization, and Techniques*. CRC Press, Boca Raton, UK.
6. Tripathi, S. L., Ganguli, S., Kumar, A., & Magradze, T. (Eds.). (2022). *Intelligent green technologies for sustainable smart cities*. John Wiley & Sons, USA, doi: 10.1002/9781119816096

6

Development of New Bioinspired Hybrid Algorithms for Parameter Modeling of Photovoltaic Panels

Souvik Ganguli*, Shilpy Goyal and Parag Nijhawan

Department of Electrical and Instrumentation Engineering, Thapar Institute of Engineering and Technology, Patiala, Punjab, India

Abstract

Three hybrid bioinspired techniques are developed in this chapter. Two popular nature-dependent methods, *viz.*, the firefly method, abbreviated as firefly algorithms (FA) and the bacterial foraging algorithm (BFA) were integrated utilizing the benefits of each. Firefly algorithms is employed for exploring the entire search domain while the three improved BFA schemes proposed by Supriyono have been utilized to search locally in the global optimization methods. Mainly, the step size of chemotaxis in bacterial foraging technique is modified with linear, quadratic, and exponential adaptations to improve upon the basic BFA technique. The bioinspired techniques not only outperform their parent methods, but they also surpass some of the most cutting-edge metaheuristic strategies. A bunch of unconstrained standard functions are taken up in this work to test the efficiency of the suggested approaches. The statistical analysis of the optimal values as well their non-parametric test also validate this fact. Further, these hybrid algorithms are also successfully used for the evaluation of two-diode model parameters using a novel objective function coined as weighted sum of square error (WSSE). The test results of the two different photovoltaic systems truly prove that the proposed method is effective in obtaining the unknown model parameters. The variation of the cost function with the iteration counts also verify the faster convergence and accuracy of the results. In future, multi-dimensional diode modeling can be considered to identify their undetermined parameters with the proposed technique. Instead of arbitrary weights, optimized weights can be assigned to bring about even better

**Corresponding author*: souvik.ganguli@thapar.edu

Tamal Roy, Suman Lata Tripathi and Souvik Ganguli (eds.) Linear and Nonlinear System Modeling, (89–103) © 2024 Scrivener Publishing LLC

results. Parameter modeling of various photovoltaic panels can also benefit from a multi-objective strategy.

Keywords: Firefly algorithms (FA), bacterial foraging algorithm (BFA), modified bacterial foraging algorithm (MBFA), hybrid firefly algorithm (HFA), bioinspired technique, two-diode model

6.1 Introduction

In bioinspired techniques, hybrid algorithms have been developed by integrating two or more algorithms to improve or enhance overall search efficiency. Many researchers have quite often tried to utilize the merits of each algorithm for the common good [1]. In this chapter, three new bioinspired hybrid propositions are presented by integrating the firefly algorithm (FA) [2] with some improved forms of bacterial foraging [3]. The bacterial foraging technique, suggested by Passino, is evolved on the basis of the foraging phenomenon of *E. Coli* bacteria that stays within the intestine of the human body. But, bacterial foraging algorithm (BFA) uses a lot of user-defined parameters, suffers from the drawback of premature convergence, and its efficacy heavily decreases with dimensions and problem complexity [4], for which an integrated approach of FA and BFA is conceptualized. A bacterial foraging system comprises four salient components, namely, chemotaxis followed by swarming, reproduction, and finally, the elimination-dispersal. Out of these four key steps of the algorithm, adaptive variation of the chemotactic step, is widely popular in the literature.

Supriyono [5] proposed one such scheme to bring about the linear, quadratic, and exponential formulation of the chemotactic step which improved the standard BFA to a considerable extent. This strategy is used to yield a modified bacterial foraging algorithm, coined as MBFA in this chapter. Thus, three global optimization methods are presented combining FA with MBFA. Firefly algorithm is used for exploration purpose while three MBFA methods are employed to execute the exploitation feature. The hybrid algorithms are referred to in this chapter as FABFA-1, FABFA-2, and FABFA-3, respectively. Both unimodal and multimodal benchmark functions of appreciable dimensions are considered to test the efficacy of these new hybrid algorithms. A sufficient number of recently developed metaheuristic algorithms are set up for a comparative study. Statistical tests, both parametric and non-parametric, are conducted to validate the experimental outcomes. The convergence curves also prove the accuracy of results and the quality of solutions.

Further, these novel hybrid propositions were utilized to identify several model specifications of commercial photovoltaic modules directly from its datasheet information [6]. In order to optimize for these unknown model parameters, the root mean square error (RMSE) [7] is often considered. The square error is equally competent to handle this problem as well. In this chapter a weightage is assigned to the sum of square error (SSE) concept to obtain the two-diode model parameters with the help of the hybrid algorithms instead of the typical square error to give an edge over it. The popular two-diode model [8] is taken up for parameter estimation of some common solar cells applied in the industries [9]. A good amount of comparison is also provided to give justification to the efficacy of the advocated methods [10–13]. Thus, the chapter efficiently develops new bioinspired methods for modeling photovoltaic panels using a non-experimental approach.

The remainder chapter flow is developed in the following way. The two-diode model equations are deliberated in Section 6.2 for the understanding of the readers. Even the cost function is also suggested in this section as well. The hybrid approaches constituting the firefly technique and the improved BFAs are deliberated in Section 6.3. The approach of work is also briefed here as well. Section 6.4 showcases the simulation outcomes, while Section 6.5 draws the salient conclusions with few indications for future propositions.

6.2 Problem Statement

A two-diode model is considered here for the study. Owing to the losses due to the recombination phenomenon taking place in the depletion part of the semiconductor, an extra diode is introduced to the single diode representation of the photovoltaic system. Thus, the prime merit of this model as compared to the single-diode system is that it gives rise to superior precision in cases of relatively lower irradiation. The equivalence for the two-diode model is shown in Figure 6.1. The current for this two-diode model is thus represented by

$$I = I_{PV} - I_{01}\left[exp\left(\frac{q(V+IR_S)}{\alpha_1 KT}\right)-1\right] - I_{02}\left[exp\left(\frac{q(V+IR_S)}{\alpha_2 KT}\right)-1\right]-\frac{V+IR_S}{R_p} \tag{6.1}$$

where 'I_{PV}' represents the photocurrent, while 'I_{01}' and 'I_{02}' typify the reverse saturation currents of the respective diodes, 'R_p' and 'R_S' denote,

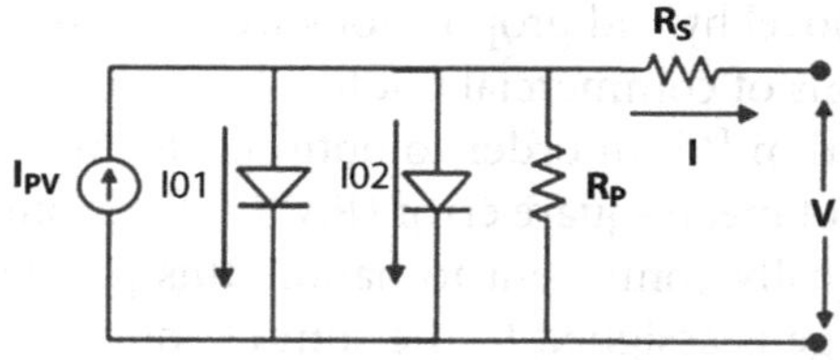

Figure 6.1 The electrical circuit representative for a two-diode model.

respectively, the resistances connected in parallel and series arrangement in the circuit, α_1 and α_2 are the diode ideality factors, 'q' replicate the charge of an electron, 'T' denotes the temperature expressed in absolute scale, and last but not the least, 'K' represents the Boltzmann constant in Equation 6.1.

At the time of an open circuit, the current becomes equal to zero, whereas $V = V_{OC}$, representing the open-circuit voltage; thus, Equation 6.1 can be modified as

$$0 = I_{PV} - I_{01}\left[exp\left(\frac{qV_{OC}}{\alpha_1 KT}\right) - 1\right] - I_{02}\left[exp\left(\frac{qV_{OC}}{\alpha_2 KT}\right) - 1\right] - \frac{V_{OC}}{R_p} \tag{6.2}$$

On solving,

$$I_{PV} = I_{01}\left[exp\left(\frac{qV_{OC}}{\alpha_1 KT}\right) - 1\right] + I_{02}\left[exp\left(\frac{qV_{OC}}{\alpha_2 KT}\right) - 1\right] + \frac{V_{OC}}{R_p} \tag{6.3}$$

Under the shorted conditions, the voltage 'V' will be zero while $I = I_{SC}$, denoting the short-circuit current. In that case, Equation 6.1 can be written as

$$I_{SC} = I_{PV} - I_{01}\left[exp\left(\frac{qI_{SC}R_S}{\alpha_1 KT}\right) - 1\right] - I_{02}\left[exp\left(\frac{qI_{SC}R_S}{\alpha_2 KT}\right) - 1\right] - \frac{I_{SC}R_S}{R_p} \tag{6.4}$$

Thus,

$$I_{PV} = I_{SC} + I_{01}\left[exp\left(\frac{qI_{SC}R_S}{\alpha_1 KT}\right) - 1\right] + I_{02}\left[exp\left(\frac{qI_{SC}R_S}{\alpha_2 KT}\right) - 1\right] + \frac{I_{SC}R_S}{R_p} \tag{6.5}$$

When maximum power condition is met, the respective voltage and current are given by $V = V_m$, and $I = I_m$ therefore Equation 6.1 can alternatively be denoted by

$$\begin{aligned} I_m = I_{PV} - I_{01}\left[exp\left(\frac{q(V_m + I_m R_S)}{\alpha_1 KT}\right) - 1\right] \\ - I_{02}\left[exp\left(\frac{q(V_m + I_m R_S)}{\alpha_2 KT}\right) - 1\right] - \frac{V_m + I_m R_S}{R_p} \end{aligned} \tag{6.6}$$

Our objective is to determine accurately the parameters of photovoltaic systems for the three important criteria mentioned above. Thus, in order to minimize the errors, the help of a metaheuristic technique is sought.

From Equation 6.3, the error value during the open-circuit condition can be calculated as follows.

$$err_{OC} = I_{01}\left[exp\left(\frac{qV_{OC}}{\alpha_1 KT}\right) - 1\right] + I_{02}\left[exp\left(\frac{qV_{OC}}{\alpha_2 KT}\right) - 1\right] + \frac{V_{OC}}{R_p} - I_{PV} \tag{6.7}$$

Similarly, the error function for the short-circuit condition can be formulated with Equation 6.5, as given below.

$$\begin{aligned} err_{SC} = I_{SC} + I_{01}\left[exp\left(\frac{qI_{SC}R_S}{\alpha_1 KT}\right) - 1\right] \\ + I_{02}\left[exp\left(\frac{qI_{SC}R_S}{\alpha_2 KT}\right) - 1\right] + \frac{I_{SC}R_S}{R_p} - I_{PV} \end{aligned} \tag{6.8}$$

Finally, Equation 6.6, depicting the maximum power point condition, is utilized to derive the last error equation denoted by

$$err_m = I_{PV} - I_{01}\left[exp\left(\frac{q(V_m + I_m R_S)}{\alpha_1 KT}\right) - 1\right] - I_{02}\left[exp\left(\frac{q(V_m + I_m R_S)}{\alpha_2 KT}\right) - 1\right] - \frac{V_m + I_m R_S}{R_p} - I_m \quad (6.9)$$

Thus, the objective function, namely the weighted sum of square error (WSSE) is framed as

$$err = w_1 \times err_{OC}^2 + w_2 \times err_{SC}^2 + w_3 \times err_m^2 \quad (6.10)$$

Arbitrary weights, mostly equal in the present chapter, are assigned to the SSE to get improved results.

6.3 Proposed Bioinspired Techniques and Methodology

The BFA and the firefly technique are quite popular in the literature and have been utilized in manifold applications [14–17]. One of the major drawbacks of the BFA is that it uses a lot of user-defined parameters. The choice of appropriate parameters thus portrays a vital role in obtaining the solutions of quality. Further, increasing the dimensionality of the problem also has a huge impact on the outcomes of the experiment. Several works on the modifications of the parent BFA technique were suggested in the literature [18–21]. Most of them were centered around improving the results of the conventional BFA technique by bringing about changes in its chemotactic step size. Supriyono proposed three different ways of varying the chemotactic step size [5]. The step size was varied adaptively with the help of either linear, quadratic, or exponential relations to improve upon the results over the original BFA method. Three hybrid combinations are thus developed integrating FA with the abovementioned adaptive BFA techniques. The hybrid algorithms so developed are coined as FABFA-1, FABFA-2, and FABFA-3, respectively, in this chapter. These algorithms are, first of all, tested on some representative high dimensional benchmark functions. Moreover, these algorithms are further utilized to find out the different parameters of the two-diode models. Quite a substantial number of methods are employed to compare with the proposed approaches. The statistical measures of the results are also reported for the test functions. In order to verify the significance of the findings, the Wilcoxon's rank-sum test [22] is also performed. The plot of the fitness function against the total number of iterations also proves better convergence speed and accuracy of the suggested techniques over the compared methods.

6.4 Simulation Results and Discussions

A small set of typical tasks was chosen to test the effectiveness of the proposed techniques. The nature of the test functions is not limited in any way, and they are typically categorized as unimodal or multimodal. To reflect the proposed techniques' diversification and intensification, five test functions from each of the unimodal and multimodal categories were chosen as representatives. A high-dimensional analysis is performed to appropriately determine the efficiency of the suggested methods. The details of the test functions considered for this work are already described in Reference [23]. Hundred decision variables are optimized using the proposed techniques. Some common details of these testbeds with respect to the function type, their search limits, and further, the ideal optimum values are given in Table 6.1.

All the test functions Func1–Func10 are to be minimized with an ideal optimum value of zero. The number of search agents for these problems is taken as 30 whereas the number of iterations is considered as 500 to solve these unconstrained optimization problems. Table 6.1 shows that these values compare favorably to those involving one hundred decision variables and fixed search parameters. All algorithms used for comparison have had their parameters set according to what is considered standard practice in the relevant literature. Given the random nature of the metaheuristic algorithm's output, we run it 20 times to draw statistically significant conclusions. Table 6.2 displays these statistical evaluations of the test results. Each method's best and worst results as well as their average and standard deviation are detailed. Table 6.2's minimum values for each column are highlighted in bold.

On most occasions, the proposed hybrid mechanisms perform better as compared to as many as seven algorithms as indicated by the bolded letters in Table 6.2. The grey wolf optimizer (GWO) and BFA techniques are exceptions on two occasions. Among the hybrids proposed, FBFA-3

Table 6.1 Benchmark test problems and their descriptions.

Nomenclature of the functions	Function names	Type of function	Lower and upper bounds	f_{min}
Func1	Sphere	Unimodal	[–100, 100]	0
Func2	Schwefel 2.22	Unimodal	[–10,10]	0
Func3	Schwefel 2.21	Unimodal	[–100,100]	0
Func4	Rosenbrock	Unimodal	[–30,30]	0
Func5	Step	Unimodal	[–100,100]	0
Func6	Rastrigin	Multimodal	[–5.12,5.12]	0
Func7	Ackley	Multimodal	[–32,32]	0
Func8	Griewank	Multimodal	[–600,600]	0
Func9	Penalized-1	Multimodal	[–50,50]	0
Func10	Penalized-2	Multimodal	[–50,50]	0

proves to be the preferred one based on the results obtained. Even the performance of FABFA-1 and FABFA-2 are better on numerous occasions as compared to the other techniques compared. To assess the implication of the outcomes, the Wilcoxon test measuring the rank-sum of the datasets [22] was performed. The p-values are calculated in this evaluation. The approximate orders of the results are reported in Table 6.3. The p-values having the numerical figure greater than 0.05 is considered to be insignificant and hence underlined. Mostly, the p-values are similar on most occasions, and hence, represented with the order only. Few distinct values near 0.05 are also indicated in this table.

It is clear from Table 6.3 that almost 95% of the results are significant. Thus, the results of the proposed bioinspired techniques are validated by the non-parametric test. To test an engineering application, the proposed methods are applied for the parameter modeling of the two-diode models of some commercial solar cells. The datasheets of the solar cells used are reported in Table 6.4 for the benefit of curious readers.

The various terms used in the Table 6.4 carry their usual significance. The population size considered for optimizing the PV cell parameters is 50, while the iteration count for this problem is taken up as 1000. Various methods are employed to find the seven unidentified parameters (I_{PV}, α_1, α_2, R_S, R_P, I_{01}, I_{02}) of the two-diode model. The parameter limits are

Table 6.2 A comparative study to assess the performance of the proposed methods with other algorithms.

Test functions	Algorithms	Minimum	Worst	Mean	Std.
Func1	FABFA-1	3.9004E−15	7.5817E−15	5.0862E−15	9.5852E−16
	FABFA-2	9.4749E−16	1.6713E−15	1.2089E−15	1.9754E−16
	FABFA-3	**6.1223E−16**	**1.0189E−15**	**7.8070E−16**	**1.2164E−16**
	FA [2]	5.92E−06	1.2E−05	8.4E−06	1.58E−06
	BFA [3]	51.66	615.5578	177.1513	134.3055
	GWO [10]	2.7E−13	2.96E−12	1.23E−12	7.68E−13
	DA [11]	823.73	5.16E+04	2.28E+04	1.13E+04
	SCA [12]	2.15E+03	2.38E+04	8.3E+03	5.95E+03
	SSA [13]	8.82E+02	2.16E+03	1.48E+03	3.26E+02
Func2	FABFA-1	5.1529E−08	7.3352E−08	6.1445E−08	5.6249E−09
	FABFA-2	2.7822E−08	3.5751E−08	3.2807E−08	2.1352E−09
	FABFA-3	**1.7457E−08**	**2.5671E−08**	**2.1505E−08**	**1.6936E−09**
	FA [2]	0.0017	0.0021	0.0019	1.0178E−04
	BFA [3]	7.1492	18.3991	13.5225	3.3672
	GWO [10]	2.22E−08	8.23E−08	4.43E−08	1.46E−08
	DA [11]	41.8502	225.7288	85.9071	38.3511
	SCA [12]	0.5664	19.3137	7.4521	5.5944
	SSA [13]	37.2278	53.2753	44.4958	4.0849
Func3	FABFA-1	4.3981E−04	6.2700E−04	5.3887E−04	4.2540E−05
	FABFA-2	3.3785E−04	5.0892E−04	4.3484E−04	4.4338E−05
	FABFA-3	**1.8406E−04**	**2.7000E−04**	**2.4045E−04**	**3.7507E−05**
	FA [2]	63.0482	96.8883	84.0101	8.6711
	BFA [3]	18.7254	29.8706	24.8083	2.8613
	GWO [10]	0.2476	2.1324	0.8602	0.5228
	DA [11]	45.2060	69.1119	56.5179	6.4487
	SCA [12]	86.7194	94.7967	90.28	1.8711
	SSA [13]	21.9577	32.3182	26.9959	3.0592

(Continued)

Table 6.2 A comparative study to assess the performance of the proposed methods with other algorithms. (*Continued*)

Test functions	Algorithms	Minimum	Worst	Mean	Std.
Func4	FABFA-1	96.2882	**98.4344**	96.5032	**0.6605**
	FABFA-2	94.8917	259.9506	155.1393	42.7433
	FABFA-3	92.8388	98.6646	**95.5091**	1.9275
	FA [2]	175.0484	701.5454	316.8462	121.0249
	BFA [3]	2.46E+03	5.17E+04	1.28E+04	1.36E+04
	GWO [10]	96.1101	98.5268	97.8319	0.7269
	DA [11]	2.5E+06	6.1E+07	2.26E+07	1.37E+07
	SCA [12]	3.31E+07	2.22E+08	1.33E+08	5.24E−07
	SSA [13]	9.6E+04	4.4E+05	1.97E+05	9.05E−04
Func5	FABFA-1	1.4981E−09	2.7304E−09	2.0071E−09	3.7457E−10
	FABFA-2	3.6656E−10	**7.0816E−10**	**5.1331E−10**	**8.5000E−11**
	FABFA-3	**3.6526E−10**	8.4021E−10	5.6333E−10	1.0698E−10
	FA [2]	5.00E−06	1.15E−05	7.83E−06	1.42E−06
	BFA [3]	54.8905	620.5757	202.8115	135.7646
	GWO [10]	8.5219	11.5055	10.0113	0.8257
	DA [11]	6.2E+03	4.77E+04	2.1E+04	1.160E+04
	SCA [12]	1.2E+03	2.3E+04	9.47E+03	7.32E+03
	SSA [13]	852.7054	2.0098E+03	1.3671E+03	306.1686
Func6	FABFA-1	1.4981E−09	2.7304E−09	2.0071E−09	3.7457E−10
	FABFA-2	2.9079E−09	5.3301E−09	3.6907E−09	5.8336E−10
	FABFA-3	**1.2225E−09**	**1.9314E−09**	**1.5003E−09**	**2.0897E−10**
	FA [2]	232.8198	543.2443	402.5087	81.14
	BFA [3]	155.1326	292.0099	207.4837	34.3790
	GWO [10]	4.1604	16.5307	6.0957	4.7768
	DA [11]	547.8203	1.06e+03	839.8859	145.6656
	SCA [12]	95.9183	404.3979	240.834	81.1203
	SSA [13]	184.4813	331.7451	244.9028	38.7924

(*Continued*)

Table 6.2 A comparative study to assess the performance of the proposed methods with other algorithms. (*Continued*)

Test functions	Algorithms	Minimum	Worst	Mean	Std.
Func7	FABFA-1	9.6031E−09	1.6878E−08	1.2470E−08	2.2456E−09
	FABFA-2	4.8868E−09	6.5092E−09	5.6100E−09	4.0067E−10
	FABFA-3	**4.2271E−09**	**5.7621E−09**	**4.6420E−09**	**3.8768E−10**
	FA [2]	4.0446E−04	2.9015	1.9619	0.5866
	BFA [3]	3.9650	6.6358	5.2797	0.7711
	GWO [10]	6.15E−08	2.2E−07	1.08E−07	3.95E−08
	DA [11]	10.0024	16.5795	13.5161	1.8384
	SCA [12]	7.5226	20.6768	18.0645	4.6932
	SSA [13]	8.113	12.5868	9.6826	1.1071
Func8	FABFA-1	1.4433E−14	2.9088E−14	2.1810E−14	3.6249E−15
	FABFA-2	6.4393E−15	1.2323E−14	8.8041E−15	1.6853E−15
	FABFA-3	**1.2212E−15**	**2.8866E−15**	**2.0706E−15**	**5.1859E−16**
	FA [2]	5.1366E−06	0.0246	0.0049	0.0077
	BFA [3]	1.4301	6.0224	3.1628	1.4185
	GWO [10]	1.8163E−03	0.0315	0.0039	0.0097
	DA [11]	18.7477	378.1121	208.9218	97.8009
	SCA [12]	15.0263	255.4609	116.0755	68.8237
	SSA [13]	7.8183	24.2851	14.8216	4.3753
Func9	FABFA-1	0.5023	5.4214	2.6576	1.5994
	FABFA-2	0.0112	1.2878	0.5626	0.3718
	FABFA-3	**4.1216E−08**	1.9402	0.4816	0.3923
	FA [2]	5.2E+09	1.26E+10	7.8E+09	2.11E+09
	BFA [3]	0.1972	**0.4625**	**0.2996**	**0.074**
	GWO [10]	2.98E+05	4.3E+06	1.36E+06	1.04E+06
	DA [11]	2.8E+07	8.25E+08	3.12E+08	2.09E+08
	SCA [12]	15.6637	66.5621	36.3138	15.5600
	SSA [13]	2.6473	4.8462	3.9961	0.5713

(*Continued*)

Table 6.2 A comparative study to assess the performance of the proposed methods with other algorithms. (*Continued*)

Test functions	Algorithms	Minimum	Worst	Mean	Std.
Func10	FABFA-1	2.8162E−04	13.0106	3.5646	3.0582
	FABFA-2	2.6768E−08	3.6085	0.9869	0.4213
	FABFA-3	**3.3638E−09**	**1.5975**	**0.3585**	**0.1228**
	FA [2]	10.1428	57.7169	14.2358	11.9196
	BFA [3]	67.4809	164.3267	104.9347	26.2083
	GWO [10]	0.1428	57.7169	11.9196	14.2358
	DA [11]	1.5E+07	1.3E+08	5.07E+07	3.25E+07
	SCA [12]	4.45E+07	1.1E+09	5.25E+08	2.9E+08
	SSA [13]	148.5	7.7E+04	9.27E+03	1.76E−04

enumerated in Table 6.5 where I_{PV} is the photovoltaic current, α_1 and α_2 denote the ideality factors of the two diodes, respectively, R_S is the resistance connected in series, R_p is the resistance connected parallelly, and I_{01} and I_{02} represent, respectively, the reverse saturation currents of the first and second diode.

The performance of the proposed techniques is not only verified with the parent techniques but also juxtaposed with other illustrious techniques like salp swarm algorithm (SSA), GWO, sine cosine algorithm (SCA), and dragonfly algorithm (DA). Table 6.6 provides the optimized parameters and cost function values of the two-diode model with various methods for the specified industrial photovoltaic systems.

It is pretty clear from Table 6.6 that the suggested methods give the least error when the comparison is carried out with the original methods as well as few of the standard heuristic approaches. Further, the convergence characteristics between the cost function and the iteration count are graphically represented in Figures 6.2 and 6.3, respectively. The evaluation of the same is carried out with GWO, DA, SCA, and SSA skipping only the parent techniques. Only the plot of the FABFA-3 is presented in the figures.

The proposed bioinspired algorithms converge faster and also accurately with respect to the different techniques mentioned. Still, DA tries to reach the best convergence. The performance of other methods used for comparison is not so well as seen from the graphs of Figures 6.2 and 6.3. Thus, the results found by the proposed techniques are found to be satisfactory as compared to the mentioned algorithms. Other modifications mentioned

Table 6.3 Approximate p-value orders of the non-parametric test given by Wilcoxon.

Benchmark functions	Hybrid algorithms	FA	BFA	GWO	DA	SCA	SSA
Func1	FABFA-1	$\simeq 10^{-8}$	$\simeq 10^{-8}$	$\simeq 10^{-8}$	$\simeq 10^{-8}$	$\simeq 10^{-8}$	$\simeq 10^{-8}$
	FABFA-2	$\simeq 10^{-8}$	$\simeq 10^{-8}$	$\simeq 10^{-8}$	$\simeq 10^{-8}$	$\simeq 10^{-8}$	$\simeq 10^{-8}$
	FABFA-3	$\simeq 10^{-8}$	$\simeq 10^{-8}$	$\simeq 10^{-8}$	$\simeq 10^{-8}$	$\simeq 10^{-8}$	$\simeq 10^{-8}$
Func2	FABFA-1	$\simeq 10^{-8}$	$\simeq 10^{-8}$	$\simeq 10^{-4}$	$\simeq 10^{-8}$	$\simeq 10^{-8}$	$\simeq 10^{-8}$
	FABFA-2	$\simeq 10^{-8}$	$\simeq 10^{-8}$	0.0010	$\simeq 10^{-8}$	$\simeq 10^{-8}$	$\simeq 10^{-8}$
	FABFA-3	$\simeq 10^{-8}$	$\simeq 10^{-8}$	$\simeq 10^{-7}$	$\simeq 10^{-8}$	$\simeq 10^{-8}$	$\simeq 10^{-8}$
Func3	FABFA-1	$\simeq 10^{-8}$	$\simeq 10^{-8}$	$\simeq 10^{-8}$	$\simeq 10^{-8}$	$\simeq 10^{-8}$	$\simeq 10^{-8}$
	FABFA-2	$\simeq 10^{-8}$	$\simeq 10^{-8}$	$\simeq 10^{-8}$	$\simeq 10^{-8}$	$\simeq 10^{-8}$	$\simeq 10^{-8}$
	FABFA-3	$\simeq 10^{-8}$	$\simeq 10^{-8}$	$\simeq 10^{-8}$	$\simeq 10^{-8}$	$\simeq 10^{-8}$	$\simeq 10^{-8}$
Func4	FABFA-1	$\simeq 10^{-8}$	$\simeq 10^{-8}$	$\simeq 10^{-5}$	$\simeq 10^{-8}$	$\simeq 10^{-8}$	$\simeq 10^{-8}$
	FABFA-2	$\simeq 10^{-8}$	$\simeq 10^{-8}$	$\simeq 10^{-4}$	$\simeq 10^{-8}$	$\simeq 10^{-8}$	$\simeq 10^{-8}$
	FABFA-3	$\simeq 10^{-8}$	$\simeq 10^{-8}$	0.0071	$\simeq 10^{-8}$	$\simeq 10^{-8}$	$\simeq 10^{-8}$
Func5	FABFA-1	$\simeq 10^{-8}$	$\simeq 10^{-8}$	$\simeq 10^{-8}$	$\simeq 10^{-8}$	$\simeq 10^{-8}$	$\simeq 10^{-8}$
	FABFA-2	$\simeq 10^{-8}$	$\simeq 10^{-8}$	$\simeq 10^{-8}$	$\simeq 10^{-8}$	$\simeq 10^{-8}$	$\simeq 10^{-8}$
	FABFA-3	$\simeq 10^{-8}$	$\simeq 10^{-8}$	$\simeq 10^{-8}$	$\simeq 10^{-8}$	$\simeq 10^{-8}$	$\simeq 10^{-8}$
Func6	FABFA-1	$\simeq 10^{-8}$	$\simeq 10^{-8}$	$\simeq 10^{-8}$	$\simeq 10^{-8}$	$\simeq 10^{-8}$	$\simeq 10^{-8}$
	FABFA-2	$\simeq 10^{-8}$	$\simeq 10^{-8}$	$\simeq 10^{-8}$	$\simeq 10^{-8}$	$\simeq 10^{-8}$	$\simeq 10^{-8}$
	FABFA-3	$\simeq 10^{-8}$	$\simeq 10^{-8}$	$\simeq 10^{-8}$	$\simeq 10^{-8}$	$\simeq 10^{-8}$	$\simeq 10^{-8}$
Func7	FABFA-1	$\simeq 10^{-8}$	$\simeq 10^{-8}$	$\simeq 10^{-8}$	$\simeq 10^{-8}$	$\simeq 10^{-8}$	$\simeq 10^{-8}$
	FABFA-2	$\simeq 10^{-8}$	$\simeq 10^{-8}$	$\simeq 10^{-8}$	$\simeq 10^{-8}$	$\simeq 10^{-8}$	$\simeq 10^{-8}$
	FABFA-3	$\simeq 10^{-8}$	$\simeq 10^{-8}$	$\simeq 10^{-8}$	$\simeq 10^{-8}$	$\simeq 10^{-8}$	$\simeq 10^{-8}$
Func8	FABFA-1	0.2785	$\simeq 10^{-8}$	$\simeq 10^{-5}$	$\simeq 10^{-8}$	$\simeq 10^{-8}$	$\simeq 10^{-8}$
	FABFA-2	0.2784	$\simeq 10^{-8}$	$\simeq 10^{-5}$	$\simeq 10^{-8}$	$\simeq 10^{-8}$	$\simeq 10^{-8}$
	FABFA-3	0.2783	$\simeq 10^{-8}$	$\simeq 10^{-5}$	$\simeq 10^{-8}$	$\simeq 10^{-8}$	$\simeq 10^{-8}$
Func9	FABFA-1	0.2085	$\simeq 10^{-7}$	$\simeq 10^{-8}$	$\simeq 10^{-8}$	$\simeq 10^{-8}$	$\simeq 10^{-8}$
	FABFA-2	$\simeq 10^{-7}$	$\simeq 10^{-8}$	0.0056	$\simeq 10^{-8}$	$\simeq 10^{-8}$	$\simeq 10^{-8}$
	FABFA-3	$\simeq 10^{-7}$	$\simeq 10^{-8}$	0.4249	$\simeq 10^{-8}$	$\simeq 10^{-8}$	$\simeq 10^{-8}$

(Continued)

Table 6.3 Approximate p-value orders of the non-parametric test given by Wilcoxon. (*Continued*)

Benchmark functions	Hybrid algorithms	FA	BFA	GWO	DA	SCA	SSA
Func10	FABFA-1	0.0106	$\simeq 10^{-8}$	$\simeq 10^{-4}$	$\simeq 10^{-8}$	$\simeq 10^{-8}$	$\simeq 10^{-8}$
	FABFA-2	$\simeq 10^{-7}$	$\simeq 10^{-8}$	$\simeq 10^{-8}$	$\simeq 10^{-8}$	$\simeq 10^{-8}$	$\simeq 10^{-8}$
	FABFA-3	$\simeq 10^{-7}$	$\simeq 10^{-8}$	$\simeq 10^{-8}$	$\simeq 10^{-8}$	$\simeq 10^{-8}$	$\simeq 10^{-8}$

Table 6.4 Information sheet for the selected photovoltaic modules [9].

Name of manufacturer	Schutten Solar	Solarworld
Model number	STM6 40-36	Pro. SW255
Cell type	Mono-crystalline	Poly-crystalline
V_m (V)	16.98	30.90
I_m (A)	1.50	8.32
V_{oc} (V)	21.02	38.00
I_{sc} (A)	1.663	8.88
Ns (number of cells)	36	60
T (°C)	51	25

Table 6.5 Range of decision variables for the two-diode model.

Parameters of the two-diode model	Set lower limit	Set upper limit
I_{PV} (A)	0.001	15
α_1	0.5	7
α_2	0.5	7
R_S (ohms)	0.001	0.5
R_P (ohms)	0.01	500
I_{01} (μA)	0	1
I_{02} (μA)	0	1

Table 6.6 Optimal two-diode model parameters using several methods for STM6 40-36 and Pro. SW255.

Model	**Algorithms**	I_{PV}	α_1	α_2	R_S	R_p	I_{01}	I_{02}	**$WSSE_{min}$**
STM6 40-36	FABFA-1	1.64	1.07	2.15	0.50	240.72	7.05E−09	4.41E−07	1.23E−05
	FABFA-2	1.65	1.08	2.11	0.50	239.87	7.07E−09	4.43E−07	1.21E−05
	FABFA-3	1.65	1.09	2.13	0.49	240.34	7.10E−09	4.44E−07	**1.15E−05**
	FA [2]	1.78	1.11	2.78	0.48	235.89	4.29E−07	2.33E−07	0.0056
	BFA [3]	1.23	1.56	2.12	0.50	234.18	7.01E−07	3.52E−07	0.0367
	GWO [10]	1.68	0.50	1.32	0.50	234.68	0.00000	2.35E−07	0.0010
	DA [11]	1.65	1.08	2.14	0.50	241.73	7.12E−09	4.46E−07	0.0001
	SCA [12]	1.79	0.50	1.25	0.50	138.37	0.00000	9.02E−07	0.0255
	SSA [13]	1.67	4.57	1.26	0.21	159.74	7.29E−07	1.02E−07	0.0003

(*Continued*)

Table 6.6 Optimal two-diode model parameters using several methods for STM6 40-36 and Pro. SW255. (*Continued*)

Model	**Algorithms**	I_{PV}	α_1	α_2	R_S	R_p	I_{01}	I_{02}	$WSSE_{min}$
Pro. SW255	FABFA-1	8.88	1.50	5.81	0.034	260.81	7.12E−07	8.15E−07	2.57E−05
	FABFA-2	8.89	1.49	5.81	0.033	262.67	7.09E−07	8.18E−07	2.05E−05
	FABFA-3	8.90	1.51	5.83	0.032	263.23	7.08E−07	8.21E−07	**1.38E−05**
	FA [2]	9.29	0.50	1.12	0.581	178.51	6.87E−07	4.45E−07	0.6428
	BFA [3]	9.35	0.52	1.14	0.672	189.21	0.0000	2.39E−07	0.7826
	GWO [10]	9.33	0.50	1.09	0.467	218.24	0.00000	1.65E−07	0.5651
	DA [11]	8.89	1.51	5.84	0.033	263.26	7.08E−07	8.20E−07	0.0002
	SCA [12]	9.70	0.50	1.11	0.500	500.00	0.00000	2.44E−07	1.0399
	SSA [13]	9.40	0.50	1.11	0.452	190.02	0.00000	2.13E−07	0.6491

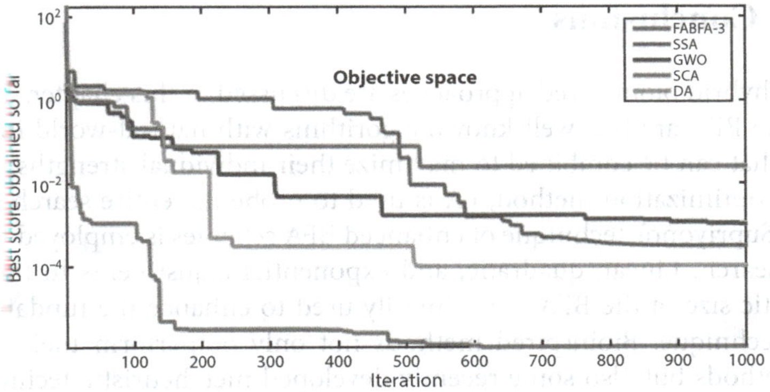

Figure 6.2 Convergence plots for DDM (Schutten Solar STM6 40-36).

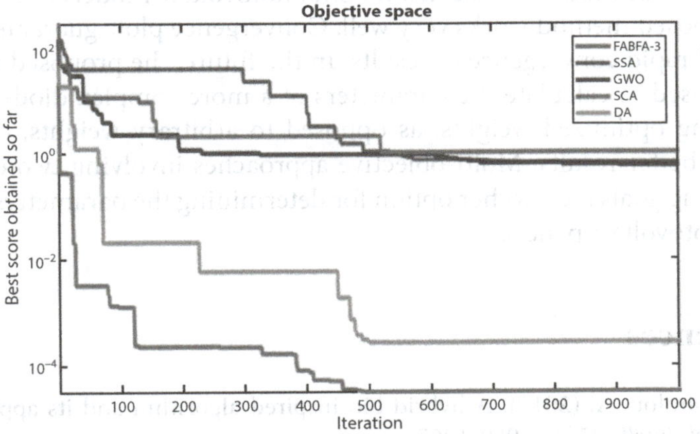

Figure 6.3 Convergence characteristics for DDM (Solarworld Pro. SW255).

below can also be tested in the future to acquire the diode model parameters for photovoltaic systems. Some typical forms of the BFA approach can be adapted for integration with FA for improved outcomes. A multi-dimensional diode model may also be constructed to produce more precise conclusions. A multi-objective strategy incorporating conflicting aims can be tried in the near future to find the parameters of the two-diode model. Since the bioinspired approaches used are stochastic methods, the results are somewhat subject to the parameter choice and the selection of termination conditions.

6.5 Conclusions

Three hybrid bioinspired approaches are discussed in this chapter. The FA and the BFA are two well-known algorithms with natural-world inspirations that can be combined to maximize their individual strengths. In the global optimization method, FA is used to probe the entire search space, while Supriyono's technique of enhanced BFA schemes is employed for the local search. Linear, quadratic, and exponential adjustments to the chemotactic size of the BFA are primarily used to enhance the fundamental BFA technique. Bioinspired methods not only outperform their ancestor methods but also some recently developed metaheuristic techniques. Using roughly ten widely used test functions, we are able to demonstrate the efficacy of our method. Moreover, the statistical methods and the non-parametric test verify the experiment results. In addition, the WSSE has proven useful with these hybrid algorithms for testing parameters in the two-diode model. Results from both photovoltaic models indicate that the proposed method works very well. Convergence plots guarantee accurate and rapid convergence of results. In the future, the proposed method can be used to calculate the parameters of a more complex diode model. Assigning optimized weights, as opposed to arbitrary weights, can lead to even better results. Multi-objective approaches involving two or more competing goals are another option for determining the parameters of various photovoltaic panels.

References

1. Hatamlou, A. (2017). A hybrid bio-inspired algorithm and its application. *Appl. Intell.*, 47(4), 1059-1067.
2. Yang, X. S. (2010). Firefly algorithm, stochastic test functions and design optimisation. *Int. J. Bio-Inspired Comput.*, 2(2), 78-84.
3. Passino, K. M. (2012). Bacterial foraging optimization. In *Innovations and Developments of Swarm Intelligence Applications* (pp. 219-234). IGI Global.
4. Das, S., Biswas, A., Dasgupta, S., & Abraham, A. (2009). Bacterial foraging optimization algorithm: theoretical foundations, analysis, and applications. In *Foundations of computational intelligence* volume 3 (pp. 23-55). Springer, Berlin, Heidelberg.
5. Supriyono, H. (2012). *Novel bacterial foraging optimisation algorithms with application to modelling and control of flexible manipulator systems* (Doctoral dissertation, University of Sheffield).

6. Biswas, P. P., Suganthan, P. N., Wu, G., & Amaratunga, G. A. (2019). Parameter estimation of solar cells using datasheet information with the application of an adaptive differential evolution algorithm. *Renew. Energy*, 132, 425-438.
7. Abd Elaziz, M., & Oliva, D. (2018). Parameter estimation of solar cells diode models by an improved opposition-based whale optimization algorithm. *Energy Convers. Manage.*, 171, 1843-1859.
8. Jordehi, A. R. (2016). Parameter estimation of solar photovoltaic (PV) cells: A review. *Renew. Sustain. Energy Rev.*, 61, 354-371.
9. Jadli, U., Thakur, P., & Shukla, R. D. (2017). A new parameter estimation method of solar photovoltaic. *IEEE J. Photovoltaics*, 8(1), 239-247.
10. Mirjalili, S., Mirjalili, S. M., & Lewis, A. (2014). Grey wolf optimizer. *Adv. Eng. Software*, 69, 46-61.
11. Mirjalili, S. (2016). Dragonfly algorithm: a new meta-heuristic optimization technique for solving single-objective, discrete, and multi-objective problems. *Neural Comput. Appl.*, 27(4), 1053-1073.
12. Mirjalili, S. (2016). SCA: a sine cosine algorithm for solving optimization problems. *Knowledge-Based Syst.*, 96, 120-133.
13. Mirjalili, S., Gandomi, A. H., Mirjalili, S. Z., Saremi, S., Faris, H., & Mirjalili, S. M. (2017). Salp Swarm Algorithm: A bio-inspired optimizer for engineering design problems. *Adv. Eng. Software*, 114, 163-191.
14. Gupta, P., & Sharma, A. K. (2019). Clustering-based Optimized HEED protocols for WSNs using bacterial foraging optimization and fuzzy logic system. *Soft Comput.*, 23(2), 507-526.
15. Chen, H., Zhang, Q., Luo, J., Xu, Y., & Zhang, X. (2020). An enhanced Bacterial Foraging Optimization and its application for training kernel extreme learning machine. *Appl. Soft Comput.*, 86, 105884.
16. Selvakumar, B., & Muneeswaran, K. (2019). Firefly algorithm based feature selection for network intrusion detection. *Comput. Secur.*, 81, 148-155.
17. Jain, L., & Katarya, R. (2019). Discover opinion leader in online social network using firefly algorithm. *Expert Syst. Appl.*, 122, 1-15.
18. Tang, K., Xiao, X., Wu, J., Yang, J., & Luo, L. (2017). An improved multilevel thresholding approach based modified bacterial foraging optimization. *Appl. Intell.*, 46(1), 214-226.
19. Chen, Y. P., Li, Y., Wang, G., Zheng, Y. F., Xu, Q., Fan, J. H., & Cui, X. T. (2017). A novel bacterial foraging optimization algorithm for feature selection. *Expert Syst. Appl.*, 83, 1-17.
20. Pang, B., Song, Y., Zhang, C., Wang, H., & Yang, R. (2019). Bacterial foraging optimization based on improved chemotaxis process and novel swarming strategy. *Appl. Intell.*, 49(4), 1283-1305.
21. Yong, Y., Ma, L., Zhao, J., & Shen, X. (2020, October). A Modified Bacterial Foraging Optimizer with Adaptive Chemotactic Step in Dynamic Search Region. In *International Conference on Intelligent Computing* (pp. 41-52). Springer, Cham.

22. Wilcoxon, F., Katti, S. K., & Wilcox, R. A. (1970). Critical values and probability levels for the Wilcoxon rank sum test and the Wilcoxon signed rank test. *Selected tables in mathematical statistics*, 1, 171-259.
23. Jamil, M., & Yang, X. S. (2013). A literature survey of benchmark functions for global optimization problems. *arXiv preprint* arXiv:1308.4008.

7

Power Quality Improvement by Using PV-Integrated DSTATCOM

Pushpanjali Shadangi[1], Sushree Diptimayee Swain[1]* and Pravat Kumar Ray[2]

[1]*Department of Electrical Engineering, O.P. Jindal University, Raigarh (C.G.), India*

[2]*Department of Electrical Engineering, NIT, Rourkela, India*

Abstract
This chapter presents a new photovoltaic system (PV)-based compensator for the elimination of power quality-related issues. In this chapter, PV-based distributed static compensator (PV-DSTATCOM) with hybrid synchronous reference frame theory (HSRF) current generation technique is proposed. The motive of this research is to integrate renewable energy source to the power grid, as the world is facing huge electric power demand worldwide. Photovoltaic system is chosen due to its popularity and decaying cost among the other renewable energy sources. Integration of PV with the conventional power grid system causes huge power quality issues and these issues can be reduced by using proper compensation techniques. To meet power quality standards under IEEE, we are using very effective control strategies based on soft computing techniques. The proposed PV-DSTATCOM is designed to compensate the harmonics and to maintain the reactive power transfer in the presence of unbalanced loads. The efficacy of this system is validated by using MATLAB simulations as well as experimental data.

Keywords: Power quality issues, PV-DSTATCOM, hybrid synchronous reference frame theory, nonlinear loads, grid stability

7.1 Introduction

As we all know, the use of non-conventional sources of energy is very much popular and also becoming necessary to meet the required energy

**Corresponding author*: Sushree.swain@opju.ac.in

Tamal Roy, Suman Lata Tripathi and Souvik Ganguli (eds.) Linear and Nonlinear System Modeling, (109–126) © 2024 Scrivener Publishing LLC

demand. The increasing consumption of energy is creating lots of stresses on the power grid. Among all the non-conventional sources, solar energy is readily available in nature and the source of energy is free. The photovoltaic (PV) (Zhang *et al.*, 2020) system is integrated with the grid (Wu *et al.*, 2021) to reduce the power demand stress on the grid. Photovoltaic system is now used as power generator for many applications that can be used as standalone PV (Verma & Singh, 2021) (Kalla, Singh, Murthy, Jain & Kant, 2018) or can be used as grid-connected photovoltaic. There are several advantages of using the PV system. When the PV system is integrated with the grid, the PV system triggers a signal in the form of current into the grid. The main benefit of grid-tied PV system is to sell surplus power to other consumers. To get the proper understanding about the photovoltaic system in DG networks (Tan & Wang, 2019) (Bernstein, Wang, Dall'Anese, Le Boudec, & Zhao, 2018) are taken into consideration. A photovoltaic device is a kind of nonlinear loads (Shadangi, Swain, Ray, & Panda, 2021) and the parameters are dependent on sunlight. An energy storage device (Yang *et al.*, 2020) is always integrated with the PV panel. The reason of connecting energy storage systems along with the PV systems is to shift peak demands when needed. Power electronics inverters (Nakata, Torii, Shiozawa, & Mototani, 2015) are also installed to convert power into AC to feed into the grid. The topology has to be opted such that the inverter should give an AC current with less harmonic distortions (Langella *et al.*, 2016). High harmonics distortion can cause poor effects on all the equipment connected to the power grid.

A good power quality (PQ) (Alam & Arya, 2020) power is free from distorsion. Deviation in voltage and current waveform from a sinusoidal one like increase/decrease in magnitude of voltage due to a change in connected load, named as sag/swell (Dongre, Mishra, & Kumar Dubey, 2020), leads to poor PQ. Similarly, nonlinear, unbalanced load, and improper wiring connections are main sources of noise. Due to such distortions, excess power consumption, malfunction of equipment, and less life span occurs. The harmonic distortions (Jo, Son, & Park, 2013) cannot be mitigated completely. It can be decreased only. The proper monitoring and control schemes of electrical system are required in order to maintain a well-powered system.

The use of nonlinear loads is increasing tremendously in high rate. So, PQ problems are increasing more significantly day by day and becoming a difficulty for both utility as well as consumers. A utility must provide better PQ to the consumer for satisfactory operation of connected devices. The mitigation of PQ problem stresses detection and localization of source and cause of distortion. An appropriate control scheme can be chosen to locate the distortions and reduce as much as possible. Here, we are using hybrid synchronous reference frame (HSRF) (Swain, Ray, & Mohanty, 2017) control technique, which is designed to correct supply power factor, eliminate harmonics, provides load balancing, and to improve the load terminal voltage.

Photovoltaic-based distributed static compensator (PV-DSTATCOM) (Agarwal, Hussain, & Singh, 2017) (Singh, Kandpal, & Hussain, 2018) provides shunt compensation (Zhai *et al.*, 2020) (de Almeida Carlos, Jacobina, & dos Santos, 2018) and injects the current at the point of common coupling (PCC). The PV-DSTATCOM, with some improvements, is able to provide fast corrections for the dynamically changing loads.

Three algorithms are stated in this chapter, namely, instantaneous reactive power theory (IRPT) (Chen, Li, Zeng, & Wang, 2018) (Wu *et al.*, 2021), modified instantaneous reactive power theory (M-IRPT) (Patnaik & Panda, 2013), and HSRF (Swain *et al.*, 2017). Two current control methods are also considered here with the performances for DSTATCOM (Kumar, Mishra, & Mekhilef, 2020) systems. These current control methods are namely, indirect current control (ICC) (Patjoshi & Mahapatra, 2013) (Meng *et al.*, 2017) (Park, Kwon, & Choi, 2017) and direct current control (DCC) (Raab, Kramer, & Ackva, 2021) (Mutschler & Marcks, 2002).

This chapter is organized as: Section 7.1 is the introduction. Section 7.2 describes the PV-based DSTATCOM model. Section 7.3 demonstrates the controller design and different control algorithms along with the ICC and DCC methods. Section 7.4 discusses the simulation results obtained. Section 7.5 illustrates the implementation of the proposed controller in laboratory-developed prototype, and the last, section 7.6, concludes the chapter.

7.2 Photovoltaic (PV)-Based DSTATCOM Model

There are several advantages of using a PV system: when the PV system is integrated with grid, it triggers a current signal into the grid. The main benefit this type of system is that they have the ability to sell surplus power to other consumers. To properly understand PV model in DG systems, PV modeling and circuit analysis of the PV module is very important. A photovoltaic device is a nonlinear device and the parameters depend essentially on sunlight and temperature.

In PV modules, there is a conversion of solar radiations into electricity that takes place. One PV array has some series and parallel combination of photo voltaic modules. Some cells are assembled together to make a module or panel. At the terminals of the PV module, power is generated if fed to the inverter of any other DC type load. The solar cell model design is totally based on how the cells are assembled together. In the solar PV module, the maximum power point (MPP) tracker allows the operation of the controller at MPP. The MPPT is associated there to maintain the pulse width of converter/inverters. There are various methods to track the MPP but among all, the perturb and observe technique is more popular.

In this chapter, PV-DSTATCOM is proposed and designed for PQ improvement. The proposed system is designed by using nonlinear modeling. The PV panel model is integrated with DSTATCOM with parallel DC capacitor along with the PI controller to get the harmonic-free power as per our requirement. It is known to us that by the use of various inductive, power electronics-based loads, we cannot completely eliminate the harmonics from the source supply but we can reduce it up to a certain extent by using appropriate control algorithm. The PV-DSTATCOM is realized using three-phase, three-wire, IGBT-based voltage source inverter with parallel DC capacitor. Control algorithm is mainly designed for supply power factor improvements, to reduce harmonics, and to maintain voltage stability at the grid. The PV-DSTATCOM provides parallel compensation and inoculate the current at common coupling point. The PV-DSTATCOM with some modifications is able to provide rapid corrections for the rapidly changing loads such as furnaces and traction. Phase lock loop (PLL) is introduced to generate firing pulse for insulated gate bipolar transistor (IGBT). The PV-integrated DSTATCOM is shown in Figure 7.1. The experimental hardware set up of the proposed method is also shown in Figure 7.2.

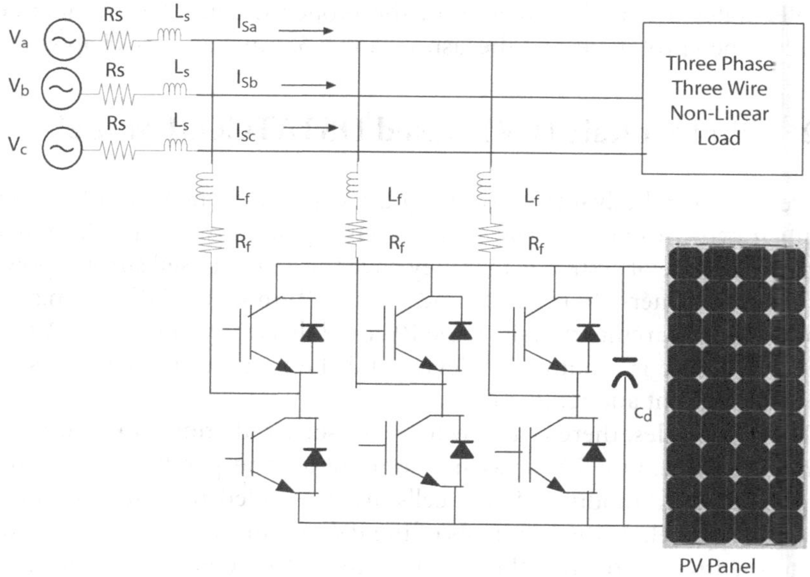

Figure 7.1 Block diagram representation of DSTATCOM.

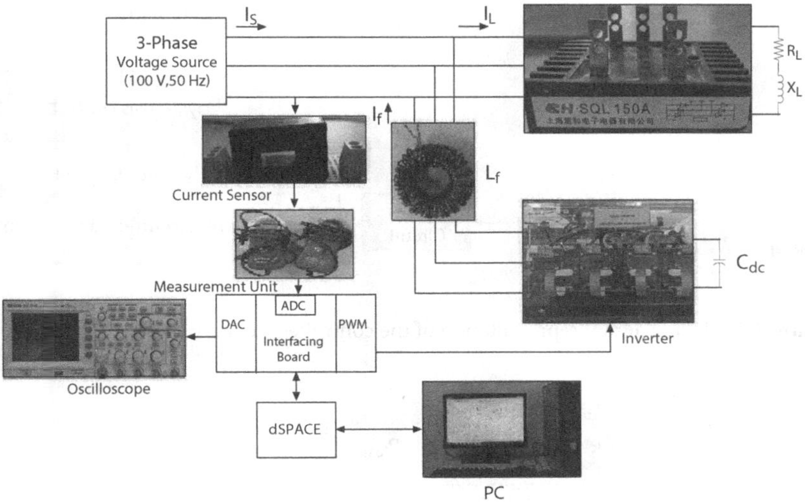

Figure 7.2 Experimental block diagram.

7.3 Controller Design and Control Algorithm

Proportional integral (PI) controller is realized here and the HSRF algorithm is discussed here in this chapter. Phase lock loop-based (PLL) synchronization technique is used, as shown in Figure 7.3. In this method, the reference current is compared with the error signal and is controlled by using a PI controller. This generated signal is fed to the inverter as a firing pulse.

7.3.1 Instantaneous Reactive Power Theory (IRPT)

In IRPT, there is a transformation of three-phase, a-b-c, to two-phase, αβ0, which takes place. Due to the use of three-phase, three-wire system consideration, zero sequence component of current gets eliminated, hence, only α–β components are considered. If the IRPT algorithm along with DSTATCOM combination is used, the voltage and currents can be tracked from the PCC. These sensed parameters are further utilized for the calculations of real and reactive powers. The block diagram of IRPT current generation scheme is shown in Figure 7.4.

The rotating frame α–β component of voltage and current are shown in Equations 7.1 and 7.2:

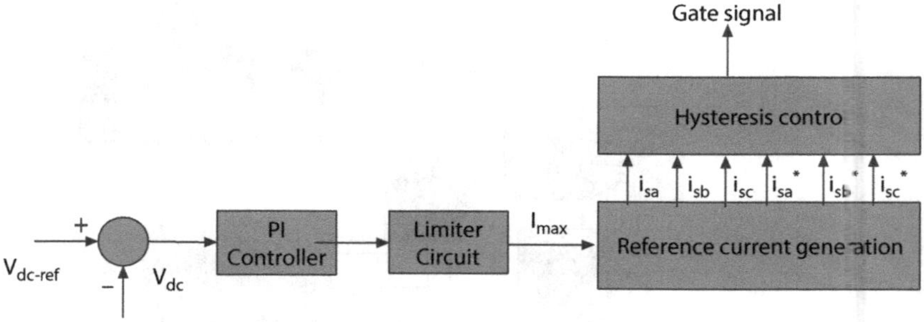

Figure 7.3 Block diagram representation of the controller block.

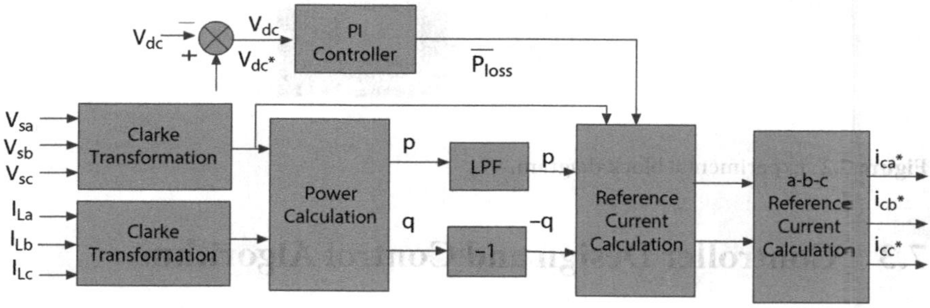

Figure 7.4 Block diagram representation of IRPT algorithm.

$$\begin{bmatrix} v_\alpha \\ v_\beta \end{bmatrix} = \sqrt{\frac{2}{3}} \begin{bmatrix} 1 & -\frac{1}{2} & -\frac{1}{2} \\ 0 & \frac{\sqrt{3}}{2} & -\frac{\sqrt{3}}{2} \end{bmatrix} \begin{bmatrix} v_{sa} \\ v_{sb} \\ v_{sc} \end{bmatrix} \tag{7.1}$$

$$\begin{bmatrix} i_\alpha \\ i_\beta \end{bmatrix} = \sqrt{\frac{2}{3}} \begin{bmatrix} 1 & -\frac{1}{2} & -\frac{1}{2} \\ 0 & \frac{\sqrt{3}}{2} & -\frac{\sqrt{3}}{2} \end{bmatrix} \begin{bmatrix} i_{La} \\ i_{Lb} \\ i_{Lc} \end{bmatrix} \tag{7.2}$$

Equations 7.1 and 7.2 can be used for the calculation of active and reactive powers. The power is shown in Equation 7.3:

$$\begin{bmatrix} p \\ q \end{bmatrix} = \begin{bmatrix} v_\alpha & v_\beta \\ -v_\beta & v_\alpha \end{bmatrix} \begin{bmatrix} i_\alpha \\ i_\beta \end{bmatrix} \tag{7.3}$$

The real and reactive powers, average, and reactive components are given by Equations 7.4 and 7.5:

$$p = \bar{p} + \tilde{p} \tag{7.4}$$

$$q = \bar{q} + \tilde{q} \tag{7.5}$$

this power is then fed to the low pass filter (LPF) and hence, the power can be obtained as in Equations 7.6 and 7.7:

$$p = \tilde{p} + p_{loss} \tag{7.6}$$

$$q = 0 \tag{7.7}$$

Reference signal are given in Equations 7.8 and 7.9:

$$\begin{bmatrix} i^*_{c\alpha} \\ i^*_{c\beta} \end{bmatrix} = \frac{1}{{v_\alpha}^2 + {v_\beta}^2} \begin{bmatrix} v_\alpha & v_\beta \\ -v_\beta & v_\alpha \end{bmatrix} \begin{bmatrix} \tilde{p} + p_{loss} \\ -q \end{bmatrix} \tag{7.8}$$

$$\begin{bmatrix} i^*_{Ca} \\ i^*_{Cb} \\ i^*_{Cc} \end{bmatrix} = \sqrt{\frac{2}{3}} \begin{bmatrix} 1 & 0 \\ -\frac{1}{2} & \frac{\sqrt{3}}{2} \\ -\frac{1}{2} & -\frac{\sqrt{3}}{2} \end{bmatrix} \begin{bmatrix} i^*_{C\alpha} \\ i^*_{C\beta} \end{bmatrix} \tag{7.9}$$

Hence, this generated reference signal is utilized for IGBT ON/OFF operation of DSTATCOM.

7.3.2 Modified Instantaneous Reactive Power Theory (MIRPT)

As there are some limitations to conventional the IRPT algorithm, hence, to cope with this limitation, we are now using this MIRPT algorithm

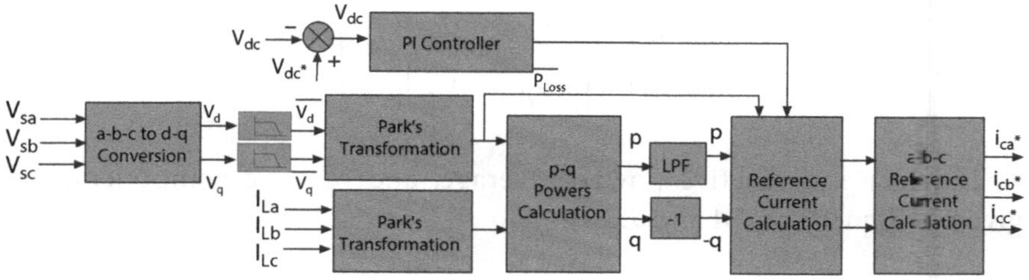

Figure 7.5 Block diagram representation of MIRPT algorithm.

for the generation of reference current. In this method, the components received after conversion from (a-b-c to d-q) form, the filter is transformed to α–β components. The equation of voltage in terms of d-q axis is shown in Equation 7.10:

$$\begin{bmatrix} v_d \\ v_q \end{bmatrix} = \sqrt{\frac{2}{3}} \begin{bmatrix} \cos\omega t & \sin\omega t \\ -\sin\omega t & \cos\omega t \end{bmatrix} \begin{bmatrix} 1 & -\frac{1}{2} & -\frac{1}{2} \\ 0 & \frac{\sqrt{3}}{2} & -\frac{\sqrt{3}}{2} \end{bmatrix} \begin{bmatrix} vsa \\ vsb \\ vsc \end{bmatrix} \tag{7.10}$$

The α–β component of voltage can be expressed, as in Equation 7.11.

$$\begin{bmatrix} v_\alpha \\ v_\beta \end{bmatrix} = \begin{bmatrix} \cos\omega t & \sin\omega t \\ -\sin\omega t & \cos\omega t \end{bmatrix} \begin{bmatrix} \bar{v}_d \\ \bar{v}_q \end{bmatrix} \tag{7.11}$$

The remaining parameters can be calculated the same as in conventional IRPT method already stated in Section 7.3.1.

Here, ω is the angular rotating speed of d-q axes rotating at synchronous speed. The d-q components of voltages received are then fed to the LPF to remove the zero sequence component. The signals received form the LPF, v_d and v_q, are then transformed to three-phase a-b-c components. All the transforms are shown by the block diagrams in Figure 7.5.

7.3.3 Hybrid Synchronous Reference Frame Theory (HSRF)

In this technique, the reference current generation takes place by combining the load voltage and source current sensing method. In this theory, the load voltage and source currents are detected simultaneously to get the harmonics. Among various control schemes available, the HSRF method is superior, it is very efficient and a reliable scheme to avoid the voltage unbalance, reduce harmonic content, and avoid voltage instability. Here, three-phase load voltages (VLa, VLb, VLc) and source currents (Isa, Isb, Isc) are transformed from a-b-c to rotating reference frame axis.

The voltages received can be further separated into two parts, DC and dynamic components, as given below in Equation 7.12:

$$\begin{aligned} V_d &= \bar{V}_d + \tilde{V}_d \\ V_d &= \bar{V}_q + \tilde{V}_q \end{aligned} \tag{7.12}$$

The load voltages are $V_{La,}$ V_{Lb}, V_{Lc}, and these are distorted and unbalanced types. If d-q-components are sent to the filter (LPF), it is subtracted from the fundamental signal. In this way, the DC components vanish and we get the harmonic signals, refer to Equation 7.13:

$$\begin{aligned} \tilde{V}_d &= V_d - \bar{V}_d \\ \tilde{V}_q &= V_q - \bar{V}_q \end{aligned} \tag{7.13}$$

Harmonic contained load voltages are given by the below Equation 7.14.

$$\begin{bmatrix} V_{Lah} \\ V_{Lbh} \\ V_{Lch} \end{bmatrix} = \begin{bmatrix} \cos\theta & \sin\theta \\ \cos(\theta - \frac{2\pi}{3}) & \sin(\theta - \frac{2\pi}{3}) \\ \cos(\theta + \frac{2\pi}{3}) & \sin(\theta + \frac{2\pi}{3}) \end{bmatrix} \begin{bmatrix} \tilde{V}_d \\ \tilde{V}_q \end{bmatrix} \tag{7.14}$$

Equations for Id and Iq are shown in Equation 7.15:

$$\begin{aligned} I_d &= \bar{I}_d + \tilde{I}_d \\ I_d &= \bar{I}_q + \tilde{I}_q \end{aligned} \tag{7.15}$$

Inverse transformation can be applied to get the harmonic contained source current as given below in Equation 7.16:

$$\begin{bmatrix} I_{sah} \\ I_{sbh} \\ I_{sch} \end{bmatrix} = \begin{bmatrix} \cos\theta & \sin\theta \\ \cos(\theta - \frac{2\pi}{3}) & \sin(\theta - \frac{2\pi}{3}) \\ \cos(\theta + \frac{2\pi}{3}) & \sin(\theta + \frac{2\pi}{3}) \end{bmatrix} \begin{bmatrix} \tilde{I}_d \\ \tilde{I}_q \end{bmatrix} \tag{7.16}$$

The reference current generation by using HSRF scheme is shown in Figure 7.6. From Figure 7.6, the error signal generated by this method is sent to the controller and harmonics can be eliminated further.

Reference compensating voltages can be expressed as Equation 7.17.

$$\begin{aligned} V_{ca}^* &= KI_{sah} - V_{Lah} + \Delta V_{caf} \\ V_{cb}^* &= KI_{sbh} - V_{Lbh} + \Delta V_{cbf} \\ V_{cc}^* &= KI_{sch} - V_{Lch} + \Delta V_{ccf} \end{aligned} \tag{7.17}$$

7.3.4 Indirect Current Control (ICC)

In the ICC method, the line currents can be acquired by getting the product of unit vectors, u_{sa}, u_{sb}, u_{sc}, and peak source current represented by Imax. The source currents are as follows in Equation 7.20. The block diagram representation of ICC method is shown in Figure 7.7.

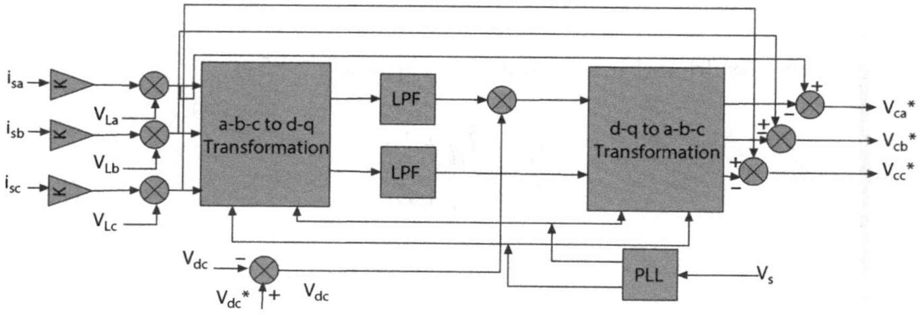

Figure 7.6 Hybrid control approach-based synchronous reference frame (HSRF) method.

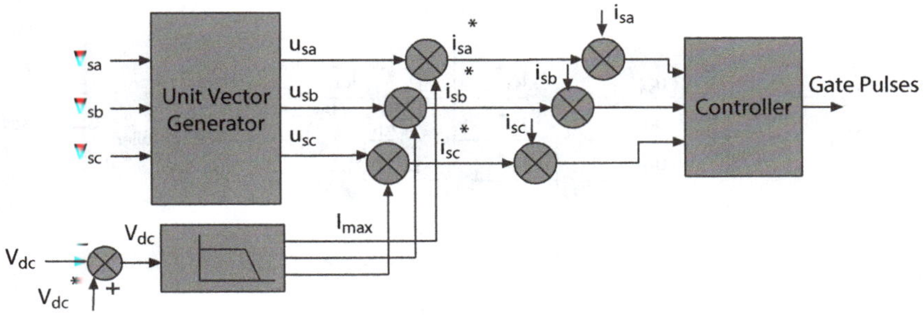

Figure 7.7 Indirect current control model.

$$V_{sa} = V_{\max} \sin \omega t$$
$$V_{sa} = V_{\max} \sin(\omega t - 120^0)$$
$$V_{sa} = V_{\max} \sin(\omega t + 120^0) \tag{7.18}$$

$$u_{sa} = u_{sa} / V_{\max}$$
$$u_{sb} = u_{sb} / V_{\max}$$
$$u_{sc} = u_{sc} / V_{\max} \tag{7.19}$$

$$I^*_{sa} = I_{\max} * u_{sa}$$
$$I^*_{sb} = I_{\max} * u_{sb}$$
$$I^*_{sc} = I_{\max} * u_{sc} \tag{7.20}$$

7.3.5 Direct Current Control (DCC)

In this method, the line current components can be obtained by subtracting the load currents from the reference source current. All the remaining equations will be the same as stated in the ICC method section 7.3.4. The block diagram representation of the DCC method is shown in Figure 7.8.

$$I^*_{ca} = I^*_{sa} - I_{La}$$
$$I^*_{cb} = I^*_{sb} - I_{Lb}$$
$$I^*_{cc} = I^*_{sc} - I_{Lc} \tag{7.21}$$

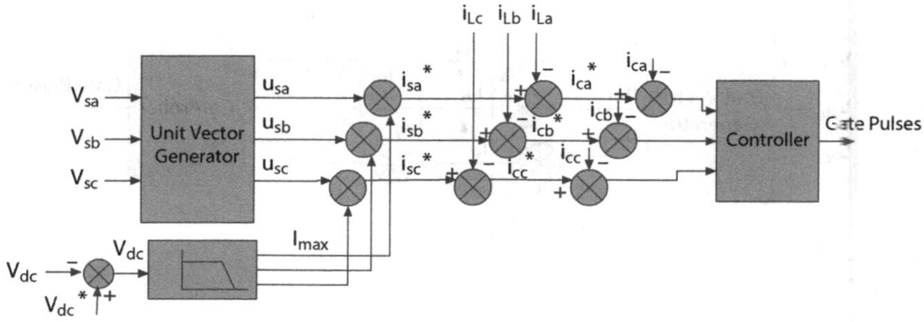

Figure 7.8 Direct current control method.

Both the methods ICC and DCC have both been analyzed to show good good total harmonic distortion (THD) spectrum but ICC shows superiority over DCC. The Table 7.1 shows the THD after compensation for both the control methods ICC and DCC for the range of order of harmonics.

7.4 Simulation Results

The PV-DSTATCOM was implemented in the MATLAB environment. The simulations were carried out for the proposed HRSF algorithm. The fast Fourier transform (FFT) was conducted to check the THD. The THD must be within the IEEE standards. Hence, it was concluded that the THD

Table 7.1 Harmonic present in source current.

Harmonic order	THD after compensation (ICC Analysis)	THD after compensation (DCC Analysis)
5	0.39	0.48
7	0.35	0.45
9	0.33	0.41
11	0.31	0.40
13	0.28	0.38
17	0.26	0.36
19	0.23	0.35

remain within the limit. For the hardware implementation verification of HSRF algorithm along with DSTATCOM, some cases were considered, as follows: (1) diode rectifier along with the resistive load, (2) diode rectifier along with resistive-inductive load, and (3) diode rectifier along with resistive-inductive-electro motive force (EMF) (R-L-E) load.

Case 1: (diode rectifier along with resistive load during steady-state condition): Figure 7.9(a), (b) and (c) reflects the MATLAB simulation results of voltage at source terminal, current at load end, and source current (post compensation) during steady-state condition with resistive load. A compensator device is inserted in the network along with pure resistive load. The current signal generated from compensator is triggered at the PCC. From Figure 7.10, we can see the THD percentage as 3.73%, which is within the standards.

Case 2: (diode rectifier along with resistive-inductive load during steady state condition): from Figure 7.9(d), we can see the THD response. For Case 2, Figure 7.10 reflects a THD of 4.23%.

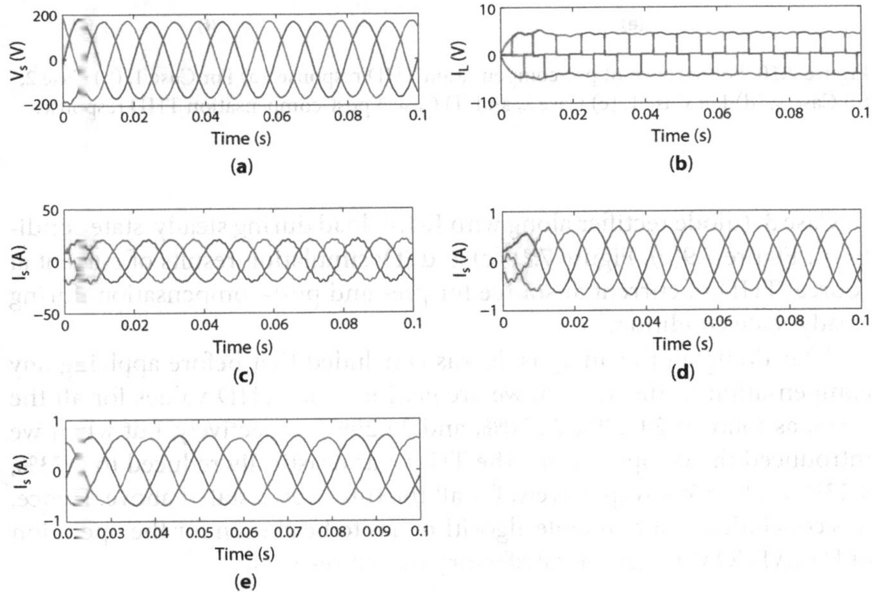

Figure 7.9 (a) Power supply voltage during steady-state response, (b) current at load without filter, (c) source current for resistive load post compensation, (d) source current for resistive-inductive load post compensation, and (e) source current for resistive-inductive-EMF load post filtration.

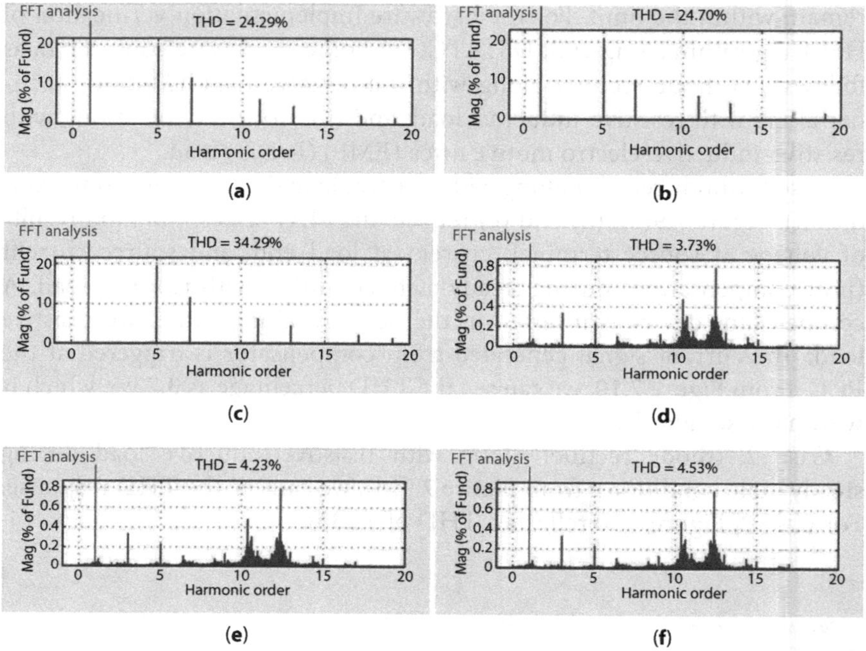

Figure 7.10 Source current pre-compensation THD response. (a) For Case 1, (b) Case 2, (c) Case 3. (d) For Case 1, (e) Case 2, and (f) Case 3 post-compensation THD response.

Case 3: (diode rectifier along with R-L-E load during steady-state condition): Figure 7.9(e), Figure 7.10(c) and (f): simulation results of current at source, THD of current at source for pre- and post-compensation during steady-state condition.

After doing all the analysis, it was concluded that before applying any compensation in the system, we are getting more THD values for all the cases, as follows: 24.29%, 24.70%, and 34.29%, respectively, but when we introduced the compensation, the THD was drastically reduced to 3.73%, 4.23%, and 4.53%, respectively, for all the three cases stated above. Hence, it is concluded that a suitable algorithm has to be chosen for the operation of DSTATCOM to get the satisfactory output response.

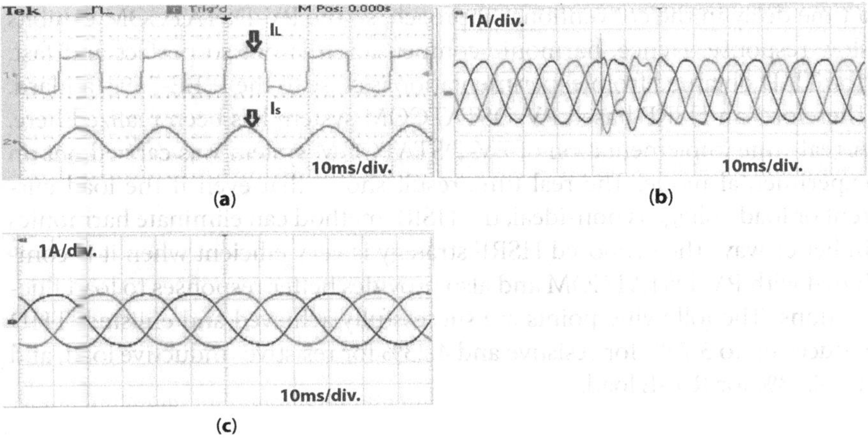

Figure 7.11 Laboratory experimental results for (a) current at load end pre compensation, current at source end post compensation for Case 1 (b) source current post compensation for Case 2, and (c) source current post filtration for Case 3.

7.5 Experimental Results

Simulations were carried out by using MATLAB 2016. To validate the working of the laboratory prototype model, experiments were performed on the hardware set up. Figure 7.2 represents the PV-based DSTATCOM model. Load dynamics were taken into account to check for various upcoming aspects.

Three cases that were taken into account: resistive load, resistive-inductive load, and R-L-E, along with diode rectifier. The supply voltage of 110 Volts, 50 Hz were selected during experiments.

The switching device SEMIKRON inverter was taken for pulse generations. Experimental results were saved by using CRO (M SO4050). Some of the figures, Figure 7.11(a), (b), and (c), respectively represents the output waveforms for all the three considered cases.

7.6 Conclusion

Implemented Photo-Voltaic Distribution Static Compensator (PV-DSTATCOM) with HSRF control algorithm as a reference voltage generation approach was proposed. The conventional control technique based PV-DSTATCOM was found to be very simple in realization and also achieve good performance when considering various load perturbations. But there is a time delay in the conventional approach, so the PV-DSTATCOM exhibits slow response. Hence, harmonic compensation is not so perfect and fast. The THD obtained from experimentation is close to the IEEE-519 standard. Therefore, an HSRF-based PV-DSTATCOM system has been realized here. A real-time implementation of PV-DSTATCOM system was carried out on experimental mode. The real-time result shows that even if the load current or load voltage is non-ideal, the HSRF method can eliminate harmonics in better way. The proposed HSRF strategy is very efficient when it is combined with PV-DSTATCOM and also provides better responses to load fluctuations. The following points are successfully achieved and enlisted: THD reduces up to 3.73% for resistive and 4.23% for resistive- inductive load, and it is 4.53% for R-L-E load.

References

Agarwal, R. K., Hussain, I., & Singh, B. (2017). Implementation of LLMF control algorithm for three-phase grid-tied spv-dstatcom system. *IEEE Trans. Ind. Electron.*, 64 (9), 7414-7424. doi:10.1109/TIE.2016.2630659

Alam, S. J., & Arya, S. R. (2020). Control of UPQC based on steady state linear Kalman filter for compensation of power quality probelms. *Chin. J. Electr. Eng.*, 6 (2), 52-65. doi:10.23919/CJEE.2020.000011

Bernstein, A., Wang, C., Dall'Anese, E., Le Boudec, J.-Y., & Zhao, C. (2018). Load flow in multiphase distribution networks: Existence, uniqueness, non-singularity and linear models. *IEEE Trans. Power Syst.*, 33 (6), 5832-5843. doi: 10.1109/TPWRS.2018.2823277

Chen, J., Li, H., Zeng, D., & Wang, G. (2018). High-impedance fault detection based on single-phase instantaneous reactive power theory. In *2018 IEEE power energy society general meeting (pesgm)* (p. 1-5). doi:10.1109/PESGM.2018.8586198

De Almeida Carlos, G. A., Jacobina, C. B., & dos Santos, E. C. (2018). Alternative breed of three-phase four-wire shunt compensators based on the cascaded transformer with single dc link. *IEEE Trans. Ind. Appl.*, 54 (3), 2492-2505. doi: 10.1109/TIA.2018.2798594

Dongre, A. A., Mishra, Y. P., & Kumar Dubey, A. (2020). Carrier PWM based capacitor supported dynamic voltage restorer for voltage sag and swell mitigation in distribution system. In *Michael faraday iet international summit 2020 (mfiis 2020)* (Vol. 2020, p. 273-278). doi:10.1049/icp.2021.1066

Jo, S.-H., Son, S., & Park, J.W. (2013). On improving distortion power quality index in distributed power grids. *IEEE Trans. Smart Grid*, 4 (1), 586-595. doi: 10.1109/TSG.2012.2222943

Kalla, U. K., Singh, B., Murthy, S. S., Jain, C., & Kant, K. (2018). Adaptive sliding mode control of standalone single-phase micro grid using hydro, wind, and solar PV array-based generation. *IEEE Trans. Smart Grid*, 9 (6), 6806-6814. doi: 10.1109/TSG.2017.2723845

Kumar, C., Mishra, M. K., & Mekhilef, S. (2020). A new voltage control strategy to improve performance of DSTATCOM in electric grid. *CES Trans. Electr. Mach. Syst.*, 4 (4), 295-302. doi:10.30941/CESTEMS.2020.00036

Langella, R., Testa, A., Meyer, J., Moller, F., Stiegler, R., & Djokic, S. Z. (2016). Experimental-based evaluation of PV inverter harmonic and interharmonics distortion due to different operating conditions. *IEEE Trans. Instrum. Meas.*, 65 (10), 2221-2233. doi: 10.1109/TIM.2016.2554378

Meng, X., Liu, Z., Liu, J., Wu, T., Wang, S., & Liu, B. (2017). A seamless transfer strategy based on indirect current control and droop control. In *2017 IEEE 3rd international future energy electronics conference and icce Asia (ifeec 2017 - ecce Asia)* (p. 1611-1616). doi:10.1109/IFEEC.2017.7992288

Mutschler, P., & Marcks, M. (2002). A direct control method for matrix converters. *IEEE Trans. Ind. Electron.*, 49 (2), 362-369. doi: 10.1109/41.993269

Nakata, A., Torii, A., Shiozawa, S., & Mototani, S. (2015). An impedance voltage compensation and vibration voltage suppression control by using inexact differential in a single-phase PWM inverter. In *Iecon 2015 - 41st annual conference of the IEEE industrial electronics society* (p.000310-000315). doi: 10.1109/IECON.2015.7392117

Park, S., Kwon, M., & Choi, S. (2017). A new anti-islanding method for indirect current control of grid connected inverter. In *2017 IEEE 3rd international future energy electronics conference and ecce Asia (ifeec2017 - ecce Asia)* (p. 1230-1234). doi: 10.1109/IFEEC.2017.7992218

Patjoshi, R. K., & Mahapatra, K. K. (2013). Performance comparison of direct and indirect current control techniques applied to a sliding mode based shunt active power filter. In *2013 annual IEEE India conference (indicon)* (p. 1-5). doi: 10.1109/INDCON.2013.6725854

Patnaik, S. S., & Panda, A. K. (2013). Real-time performance analysis and comparison of various control schemes for particle swarm optimization-based shunt active power filters. *Int. J. Electr. Power Energy Syst.*, 52, 185–197.

Raab, S., Kramer, A., & Ackva, A. (2021). Determination of RMS current load on the dc-link capacitor of voltage source converters using direct current control. *IEEE Trans. Power Electron.*, 36 (1), 968-977. doi:10.1109/TPEL.2020.3000243

Shadangi, P., Swain, S. D., Ray, P. K., & Panda, G. (2021). Experimental verification of dstatcom for various non-linear load. *Int. J. Emerging Electr. Power Syst.*.

Singh, B., Kandpal, M., & Hussain, I. (2018). Control of grid tied smart PV- dstatcom system using an adaptive technique. *IEEE Trans. Smart Grid*, 9 (5), 3986-3993. doi: 10.1109/TSG.2016.2645600

Swain, S. D., Ray, P. K., & Mohanty, K. B. (2017). Improvement of power quality using a robust hybrid series active power filter. *IEEE Trans. Power Electron.*, 32 (5), 3490-3498. doi: 10.1109/TPEL.2016.2586525

Tan, Y., & Wang, Z. (2019). Incorporating unbalanced operation constraints of three-phase distributed generation. *IEEE Trans. Power Syst.*, 34 (3), 2449-2452. doi: 10.1109/TPWRS.2019.2895559

Verma, A., & Singh, B. (2021). CAPSA based control for power quality correction in PV array integrated EVCS operating in standalone and grid connected modes. *IEEE Trans. Ind. Appl.*, 57 (2)1789-1800. doi: 10.1109/TIA.2020.3041812

Wu, F., Yang, B., Hu, A., Zhang, Y., Ge, W., Ni, L., Zha, Y. (2021).Inertia and damping analysis of grid-tied photovoltaic power generation system with dc voltage droop control. *IEEE Access*, 9, 38411-38418. doi: 10.1109/ACCESS.2021.3059687

Yang, H., Shen, W., Yu, Q., Liu, J., Jiang, Y., Ackom, E., & Dong, Z. Y. (2020). Coordinated demand response of rail transit load and energy storage system considering driving comfort. *CSEE J. Power Energy Syst.*, 6 (4), 749-759. doi: 10.17775/CSEEJPES.2020.02590

Zhai, H., Zhou, F., Zhu, C., Yi, H., Wang, Z., Tao, R., & Wei, T. (2020). An optimal compensation method of shunt active power filters for system- wide voltage quality improvement. *IEEE Trans. Ind. Electron.*, 67 (2), 1270-1281. doi: 10.1109/TIE.2019.2899561

Zhang, J., Liu, Y., Ding, K., Feng, L., Hamelmann, F. U., & Chen, X. (2020). Model parameter analysis of cracked photovoltaic module under outdoor conditions. In *2020 47th IEEE photovoltaic specialists conference (PVSC)* (p. 2).

8

Modeling and Simulation of Current Transformer to Study Its Behaviors in Different Conditions

Aveek Chattopadhyaya[1]* and Surajit Chattopadhyay[2]

[1]Department of Electrical Engineering, GNIT, Panihati, Kolkata, India
[2]Department of Electrical Engineering, GaniKhan C. Institute of Engineering and Technology, Malda, W.B., India

Abstract
Current transformer (CT) is a pertinent device in electrical systems. This chapter aims to discuss modeling and simulation of current transformer to study its behaviors. For this purpose, CT secondary signal is considered. Applying wavelet transform (WT) and fast Fourier transform (FFT), different behaviors of CT are analyzed. Wavelet-transform-based multi-resolution analysis (MRA) and root mean square (RMS) are also used to ascertain the saturation of CT and to calculate the exact primary current during saturation.

Keywords: FFT, WT, DWT, RMS, interpolation, extrapolation

8.1 Introduction

For protection and measuring purposes, Current transformer (CT) is used. Current transformer is widely used in electrical systems for smooth and reliable operation of the electrical system. For this purpose, proper modeling, simulation, and behaviors of CT in different conditions are utmost required things. Otherwise, maloperation of CT can affect the performance of the system. In Reference [1], computer simulation is discussed to show the saturation and as a repercussion the deformation of the CT secondary

**Corresponding author*: chatterjee58@gmail.com

Tamal Roy, Suman Lata Tripathi and Souvik Ganguli (eds.) Linear and Nonlinear System Modeling, (127–14?) © 2024 Scrivener Publishing LLC due to direct current (DC) component in the CT primary side. In this condition, core hysteresis and the effect of residual flux on the CT performance were also investigated. In Reference [2], saturation of CT was detected in transient conditions using sampled value of CT secondary current [2]. Y. C. Kang *et al.* proposed first, second, and third difference-based functions of CT secondary waveform to detect CT saturation [3]. In this analysis, the first difference of the CT secondary signal was used to identify the initiation of saturation. The above discussed method was tested for 100:5 CT. Negative and positive sequence impedance-based protection scheme is developed throughout saturation and ratio discrepancy [4]. The developed algorithm was verified in digital platform. N. Villamagna *et al.* proposed component (symmetrical)-based saturation detection technique in different fault conditions, where faults were created in EMTP/ATP simulation platform [5]. A digital signal processor (TMS3206701) was used to test the CT saturation detection algorithm, where the difference (third) of secondary waveform was considered for this purpose [6]. A. Hooshyar *et al.* proposed a method to calculate primary component of fault signal in existence of DC component and saturation [7]. The method is based on the look up table of fault currents and least square errors. Waveform processing and artificial intelligence (AI)-based method proposed by researchers to identify and rectify saturation for smooth operation about electrical systems [8–9]. In this analysis, signal processing such as wave transform (WT) was used to identify saturation and AI-like artificial neural network was also used to correct saturation. A. Chattopadhyaya *et al.* proposed different signal processing and signal processing-based statistical parameter analysis to assess the distorted signal in power systems [10–17]. In Reference [18], authors proposed a statistical criterion-based technique to detect CT saturation and reconstruction of CT distorted signal during saturation. Here, authors also discussed the statistical methods to discriminate fault current from inrush current. Monitoring the flux density, a method was developed to identify saturation [19].

The chapter aims to present the behaviors of CT in non-identical environments and its signal estimation on those conditions by modeling and simulation of CT. Knowledge of behaviors of CT is very essential for application and design of a CT. In this chapter, CT behaviors will also be discussed considering CT secondary signal and CT core flux. This assessment is achieved by FFT and statistical parameters based on discrete wavelet transform (DWT). This chapter will also discuss to calculate CT primary current under CT saturated conditions. Current transformer (CT) saturation for different reasons is also considered here. Here, modeling and circuit simulation of CT is discussed, which is followed by behaviors analysis, detection of CT saturation by signal processing-based statistical parameter estimations, and multi-resolution analysis (MRA)-based calculation of exact CT primary current during CT saturated conditions.

8.2 Simulation Circuit of Current Transformer

Here, the performance of CT was analyzed using the simulation circuit which is depicted in the figure below (Figure 8.1). A block diagram of a simulated network is limned in Figure 8.2. In this analysis, 120-kV network voltage is considered. 2000/5-Amp, 25-VA current transformer is used for total analysis. The mentioned CT has one primary turn and 400 secondary turns. The CT secondary is shorted by 1-ohm resistance. On the CT secondary side, a voltage measuring component is connected for measuring a secondary voltage (V2) of CT, which is proportional to the primary current (I1) of CT under normal condition. In this CT, the core flux knee point value is 10 p.u.

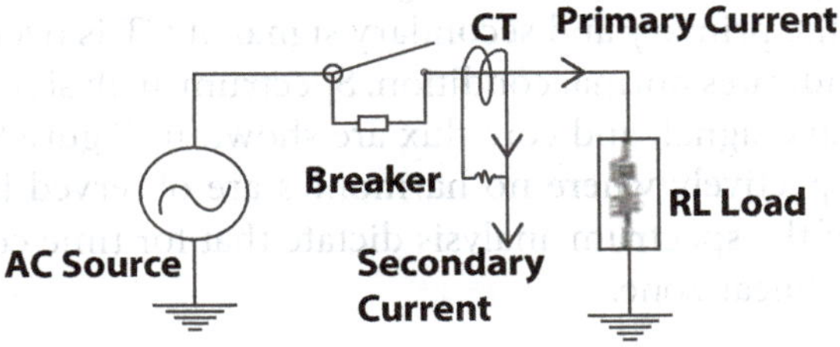

Figure 8.1 Simulation circuit of current transformer (CT).

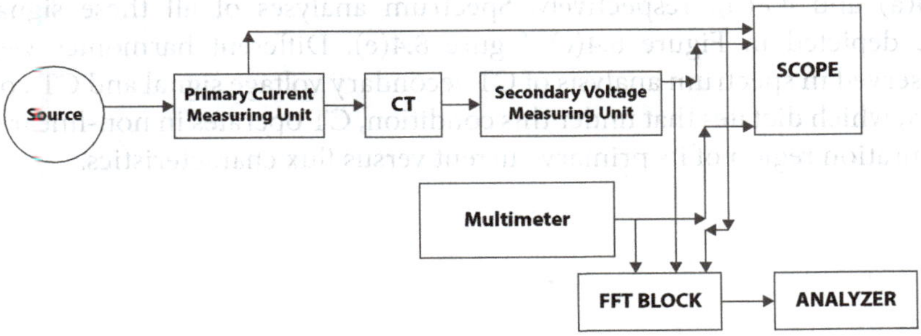

Figure 8.2 Simulation circuit in block diagram.

8.3 Effects of CT Performance Due to Variation of Circuit Time Constants

In this chapter, the CT performance was analyzed in different time constants scenario. In high voltage transmission, the line reactance is very much higher than resistance and in distribution network, the resistance is very much higher than reactance. For this reason, the performance analysis of CT is very much needed indeed for system reliability. Fast Fourier transform (FFT)-based spectrum analysis of CT primary current, secondary voltage, and CT core flux were done for this purpose.

Case Study I:
Time constant (L/R) of the circuit is 3
In Case study I, the CT performance was observed when time constant (L/R) of the circuit is 3. Fast Fourier transform-based spectrum analysis was done on the primary and secondary signal of CT and CT core flux under this condition (time constant is 3). In spectrum analysis, each frequency component consists of first positive peak (FPP), second positive peak (SPP), and negative peak (NP). Figure 8.3(a) and 8.3(b) depicts the CT primary signal and secondary signal, and core flux, respectively. In Figure 8.3(a), the primary and secondary signal of CT is overlapping each other, which indicates normal condition. Spectrum analysis of CT primary signal, secondary signal, and core flux are shown in Figure 8.3(c), 8.3(d), and 8.3(e), respectively where no harmonics are observed in this condition. Results of the spectrum analysis dictate that for time constant 3, the CT operates in linear zone.

CASE STUDY II:
Time constant (L/R) of the circuit is 0.1
The CT primary and secondary signal, and core flux are shown in Figure 8.4(a) and 4.8(b), respectively. Spectrum analyses of all these signals are depicted in Figure 8.4(c)–Figure 8.4(e). Different harmonics were observed in spectrum analysis of CT secondary voltage signal and CT core flux, which dictates that under this condition, CT operates in non-linear or saturation region of its primary current versus flux characteristics.

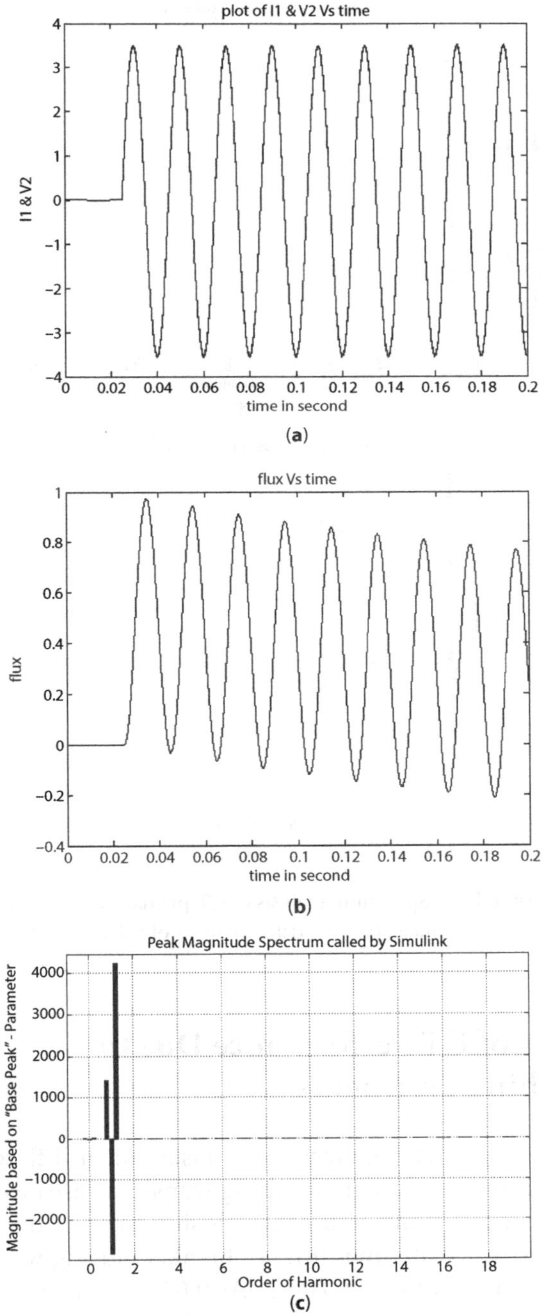

Figure 8.3 (a) CT primary signal and secondary signal. (b) Core flux of CT. *(Continued)*

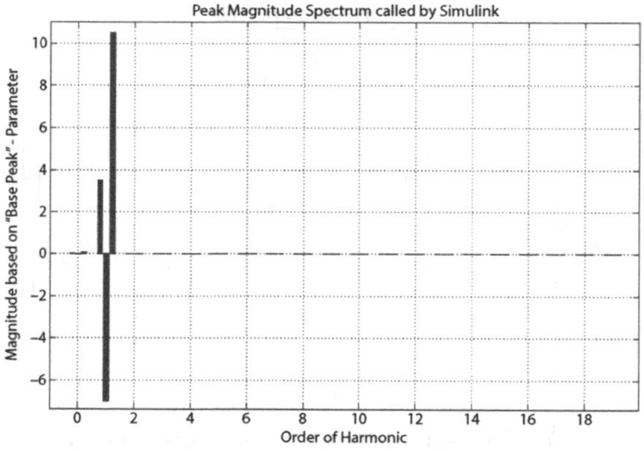

(d)

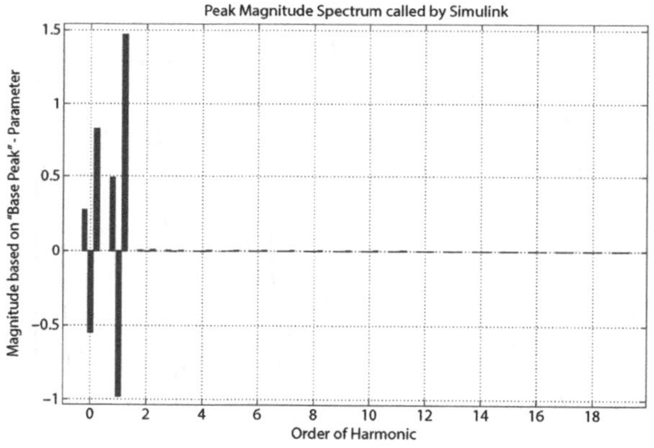

(e)

Figure 8.3 (Continued) (c) Spectrum analysis of CT primary current. (d) Spectrum analysis of CT secondary voltage. (e) Spectrum analysis of CT core flux.

8.4 Effects of CT Performance Due to Switching Transients

In this chapter, CT performances were observed in different switching transient conditions. Different switching transient cases were created by varying the switching time of breaker closing time in the primary side of the CT. Here, two switching times of breaker closing were considered. One is 0.025 s and another is 0.02 s. At 0.025 s, the voltage across the breaker is in peak position, and at 0.02 s, the voltage across the breaker

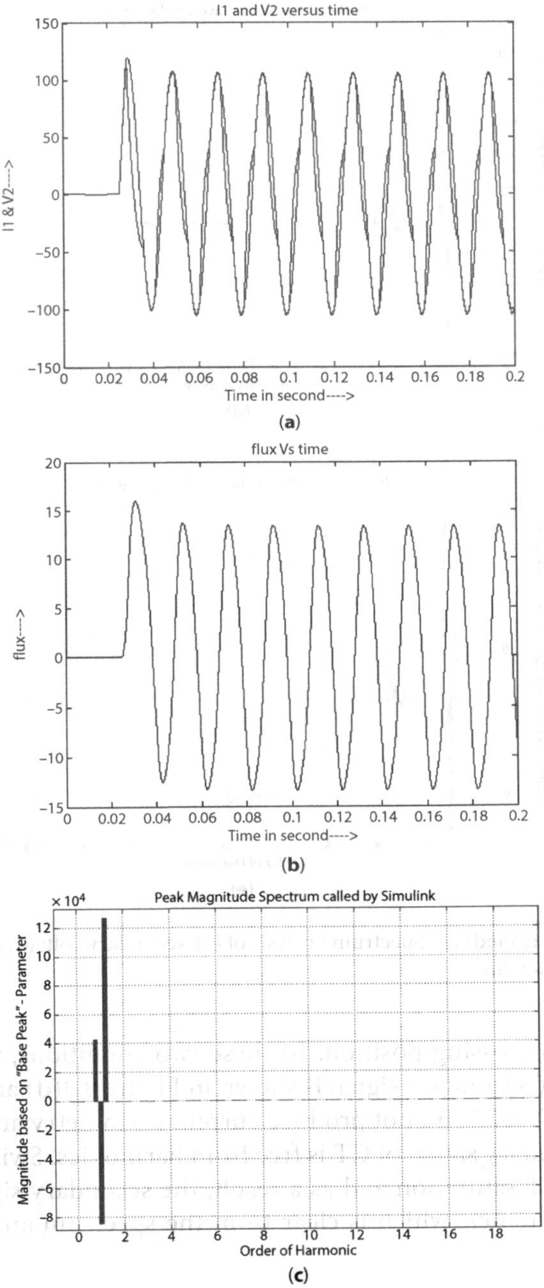

Figure 8.4 (a) CT primary signal and secondary signal. (b) Core flux of CT. (c) Spectrum analysis of CT primary current. *(Continued)*

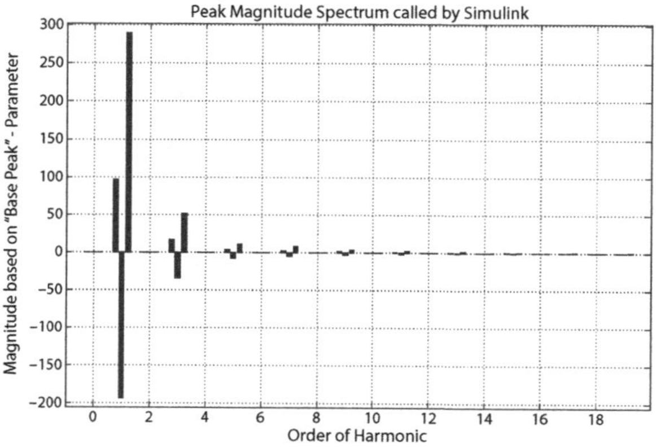

(d)

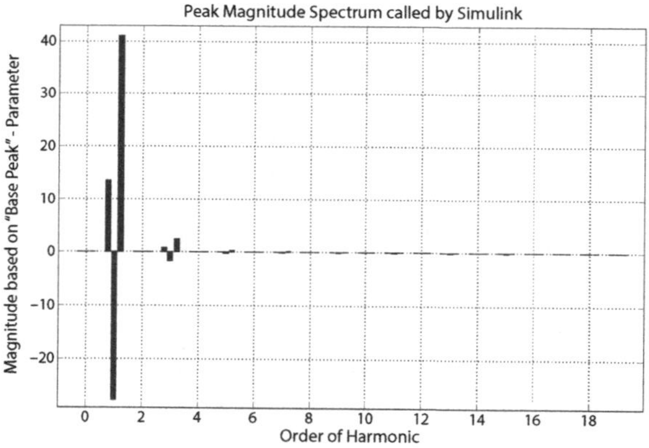

(e)

Figure 8.4 **(Continued)** (d) Spectrum analysis of CT secondary voltage. (e) Spectrum analysis of CT core flux.

is in the zero crossing position. In these two conditions, the spectrum analysis of CT secondary signal is shown in Figure 8.3(d) and Figure 8.5. Switching at 0.025 s did not produce current asymmetry and as a consequence, secondary signal of CT is free from harmonics. Switching at 0.02 s produced CT saturation and as a result, the secondary signal contains different harmonics, which is clear from the spectrum analysis (Figure 8.5).

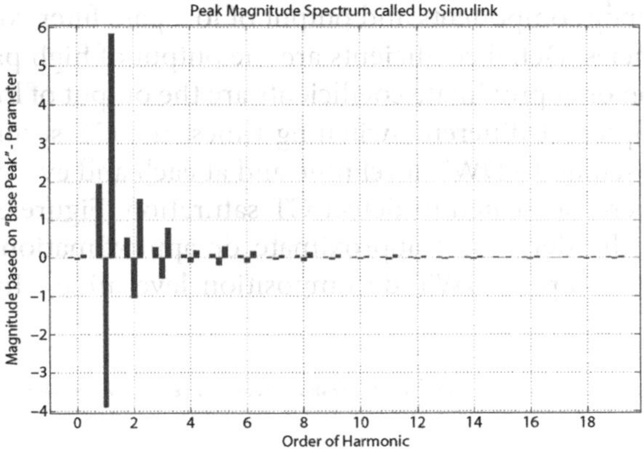

Figure 8.5 Spectrum anqalysis of CT secondary voltage (at breaker closing time is 0.02 second).

8.5 DWT-Based Skewness Analysis for Assessment of CT Saturation Due to Switching Transients

Four switching times of primary side breaker of CT were considered for this purpose. Switching times are 1.25/50 (0.025) s, 1.06/50 (0.0212) s, 1.12/50 (0.0224) s, and 1/50 (0.02) s. Among four switching times, switching at 1.25/50 (0.025) s produced no asymmetry in the CT primary current, and as result, no saturation or distortion in the CT secondary signal was found. The reason behind is that at 0.025 s, the voltage across the breaker is in peak condition. But the other three switching times produced current asymmetry, and as a result, CT saturation and distortion in the CT secondary signal were found. In this chapter, distortion of the CT secondary signal due to CT saturation for different switching times were analyzed by DWT-based skewness analysis.

Discrete Wavelet Transform (DWT)

Wavelet transform (WT) is used to get time-frequency information from a non-stationery waveform. It is categorized as *continuous wavelet transform (CWT)* and *discrete wavelet transform (DWT)*. The drawbacks of CWT are that it generates a lot of data and many complex calculations are required for CWT. Discrete wavelet transform is much faster with respect to CWT. In DWT, the signal is decomposed into two sets in each and every DWT decomposition level. In each DWT decomposition level, the signal passes through two filters (high and low pass). High pass filter output contains higher frequency components and output of low pass filter contains lower order frequencies. Detail coefficients are the output of high pass filter and approximation or approximate coefficients are the output of low pass filter.

In this chapter, at different switching times, the CT secondary signal was decomposed up to DWT level nine and at each and every level, skewness values were calculated to detect CT saturation. Figure 8.6 to Figure 8.9 delineate the detail and approximate or approximation coefficients skewness values (up to DWT decomposition level nine). In Figure 8.9,

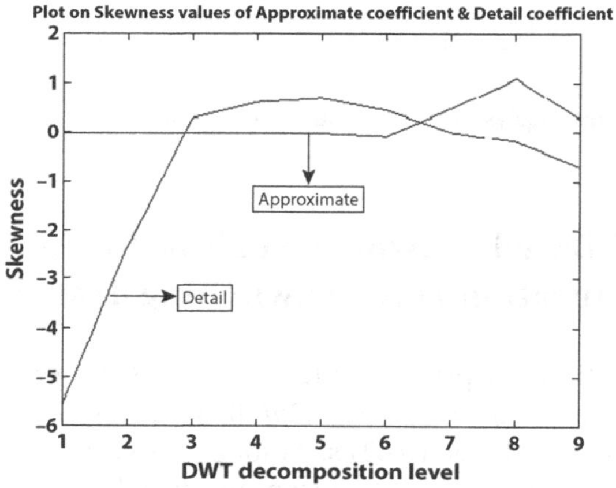

Figure 8.6 Skewness values at switching time 1/50 s.

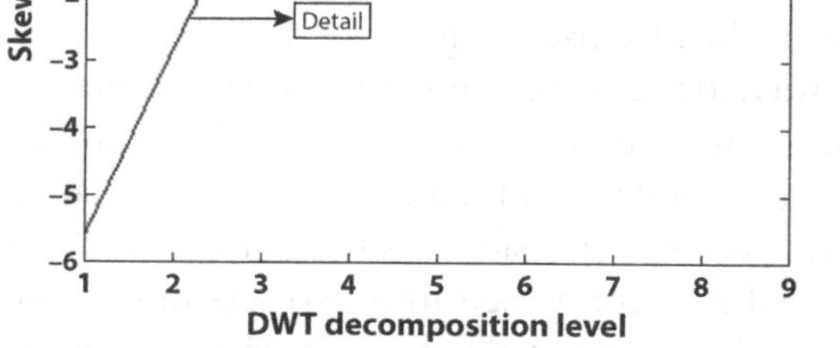

Figure 8.7 Skewness values at 1.12/50 s.

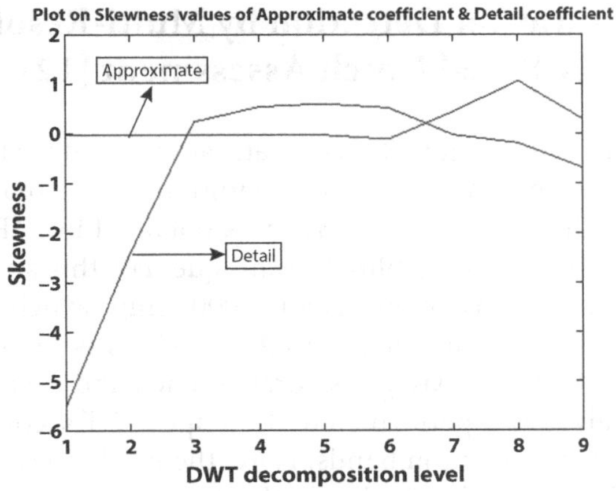

Figure 3.8 Skewness at 1.06/50 s.

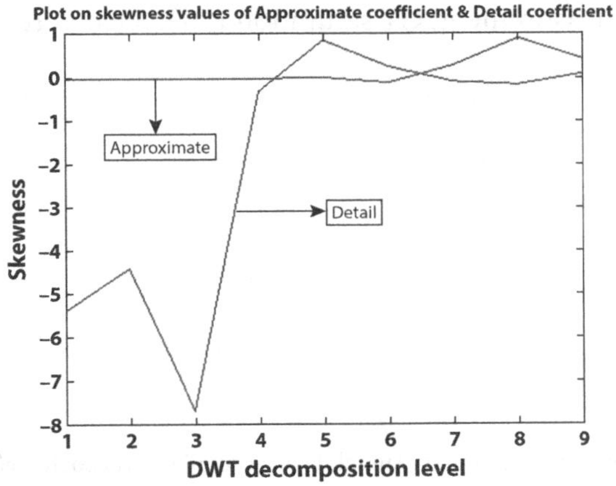

Figure 8.9 Skewness at switching time 1.25/50 s.

the skewness values of detail coefficients are totally different from the above limned figures. Another important observation is that in normal condition (switching time 1.25/60), the intersecting point between approximation and detail coefficients skewness values is between level four (4) and five (5) (in DWT level), but for otherdelineated figures, this point is in between DWT decomposition level two (2) and three (3). From these skewness values, the CT saturation or distortion of CT secondary signal can be assessed easily.

8.6 CT Saturation Detection by Multi-Resolution Analysis-Based Notch Assessment [12]

Using the main circuit (Figure 8.1), CT saturation was detected by MRA of secondary waveform and at saturated condition, actual *root mean square (RMS)* value of the primary waveform was obtained by MRA and 'interpolation' and 'extrapolation (spline)' technique. For this analysis, two CT primary current are considered. One is 1000 Amp, which is the normal condition current, and another is 25 kAmp, which is high current. Due to this, the 25-kAmp CT core got saturated. Each and every time, DWT-based MRA analysis was performed to decompose CT secondary signal in different frequency resolution bands. Here, the mother wavelet is considered as 'dB4'. Notch was observed in DWT detail level coefficients for high primary current (25kAmp) condition. Measuring this notch, CT saturation is detected. CT primary current, corresponding DWT detail level one coefficients of CT secondary signal are delineated in Figures 8.10 and 8.11. By this notch measurement, CT saturation is detected here.

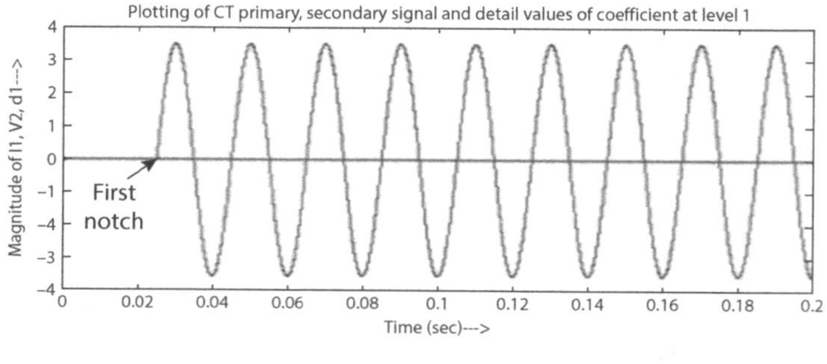

Figure 8.10 Primary and secondary signal, DWT (detail level1) coefficient.

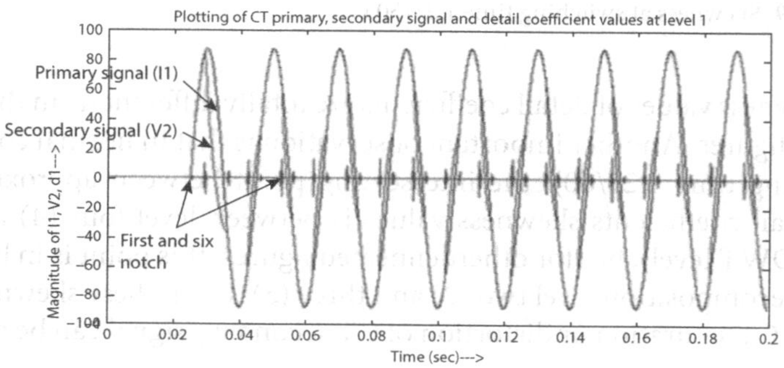

Figure 8.11 Primary, secondary signal, DWT (detail level1) coefficient.

For 1000-Amp CT primary current, at DWT detail level one (1) coefficients, no notches were observed but the maximum value of 0.00002351 (which is almost zero) is considered for analysis. Due to flow of high primary current (25 kAmp), the starting notches positive peaks are inflated with respect to normal condition. For analysis of saturation, these values were used. From Table 8.1 results, it is clear that CT saturation in each cycle can be detected by measuring the peak values of starting and ending notches. Table 8.1 is used to delineate the magnitude of starting and ending notches based on which CT saturation was detected.

Table 8.1 Starting and ending notches for different primary current.

Known primary current	Starting notch number	Positive peak value (starting notch)	No. of ending notch	Positive peak value (end notch)
1000 Amp	1	0.00002351	1	0.00002351
	2	0.00002351	2	0.00002351
	3	0.00002351	3	0.00002351
	4	0.00002351	4	0.00002351
	5	0.00002351	5	0.00002351
	6	0.00002351	6	0.00002351
	7	0.00002351	7	0.00002351
	8	0.00002351	8	0.00002351
25,000 Amp	1	9.505	1	4.281
	2	11.74	2	1.95
	3	12.56	3	3.723
	4	12.61	4	3.335
	5	8.281	5	3.595
	6	12.95	6	3.786
	7	13.33	7	4.518
	8	6.307	8	4.783

8.7 CT Primary Current Assessment During CT Saturation [12]

Exact RMS value of CT primary current is very much needed to protect from maloperation of relay. The RMS value approximation coefficients at level 1 (one) was taken here to assess the actual value of CT primary current at different conditions. For the different known values of primary currents, the RMS value of approximation level one coefficients were calculated, which is limned in Figure 8.12.Using these values, the CT primary current RMS values at different conditions were measured by extrapolation or interpolation techniques, where the results came out wildly positive. Table 8.2 delineates the proposed methods to assess CT primary currents.

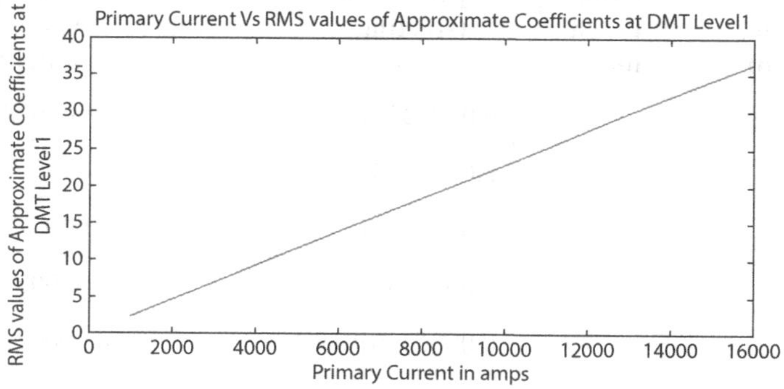

Figure 8.12 Different primary currents versus RMS value of approximate coefficients.

Table 8.2 Assessment of CT primary currents and error.

Original primary current (Amps)	7078	14,050	21,000	26,000	27,000	29,000
RMS value at DWT level one (1) (of approximation coefficients)	16.23	32.06	45.70	57.39	59.40	63.01
Calculated primary current (Amps)	7000	14,044	20,961	25,944	26,887	28,582
Percentage error	1.16%	0.042%	0.185%	0.215%	0.4185%	1.44%

8.8 Conclusion

In this chapter, circuit simulation and behaviors of CT were discussed by signal processing-based techniques. Saturation of CT was discussed in different conditions by signal assessment of CT secondary side. The CT saturation due to switching transient and high primary current was also discussed. The presence of different harmonics in CT saturated conditions was analyzed by FFT and DWT-based statistical parameter analysis. The MRA-based RMS technique was also discussed to measure the actual primary current (RMS) during CT saturation. All these things are very important for proper protection of electrical systems.

References

1. O'Kelly, D. (1992). Calculation of the transient performance of protective current transformers including core hysteresis. *IEE Proceedings-c*, 139(5), 455-460.https://10.1049/ip-c.1992.0063.
2. Evans, F. J., Wells, G. (1970). Use of a sampling scheme to detect transient saturation in protective current transformers. *IEEE Trans. Instrum. Meas.*, 19(3), 144-147. https://10.1109/TIM.1970.4313884.
3. Kang, Y. C., Ok, H., Kang, S. H., Crossley, P. A. (2004). Design and evaluation of an algorithm for detecting current transformersaturation. *IEE Proc.-Gener. Transm. Distrib.*, 151(1), 27-35. https://10.1049/ip-gtd:20030982.
4. Sachdev,M. S., Sidhu,T. S., Gill, H. S. (2000). A busbar protection technique and its performance during CT saturation and CT ratio-mismatch. *IEEE Trans. Power Delivery*, 15(3), 895-901. https:// 10.1109/61.871349.
5. Villamagna, N., Crossley, P. A. (2006). A CT saturation detection algorithm using symmetrical components for current differential protection. *IEEE Trans. Power Deliv.*, 21(1), 38-45. https://10.1109/TPWRD.2005.848654.
6. Kang, Y. C., Ok, S. H., Kang, S. H. (2004). A CT saturation detection algorithm. *IEEE Trans. Power Deliv.*, 19(1), 78-85. https://10.1109/TPWRD.2003.820200.
7. Hooshyar, A., Pasand, M. S. (2012). Accurate measurement of fault currents contaminated with decaying DC offset and CT saturation. *IEEE Trans. Power Deliv.*, 27(2), 773-783. https://10.1109/TPWRD.2011.2176965.
8. Hong, Y. Y., Chang-Chian, P. C. (2008). Detection and correction of distorted current transformer current using wavelet transform and artificial intelligence. *IET Gener. Transm. Distrib.*, 2(4), 566–575. https://10.1049/iet-gtd:20070383.
9. Yu, D. C., Cummins, J. C., Wang, Z., Yoon, H. Z., Kojovic, L. A. (2001). Correction of Current Transformer sistorted Secondary currents due to saturation using Artificial Neural Networks. *IEEE Trans. Power Deliv.*, 16(2), 189-194. https://10.1109/61.915481.

10. Chattopadhyaya, A., Chattopadhyay, S., Sengupta, S. (2013, February 26-27). Signal assessment for detection of saturation of a Current Transformer under switching transient condition by Wavelet Transform based skewness and kurtosis analysis. *Second National Conference on Instrumentation & Control, NATCONIC* – 2013, Kolkata, West Bengal, India.
11. Chattopadhyaya, A., Banerjee, H., Chattopadhyay, S., Sengupta, S. (2013). Assessment of CT saturation Caused By Switching Transient. *Int. J. Electr. Electron. Comput. Sci. Eng.*, 2(2), 57-61.
12. Chattopadhyaya, A., Chattopadhyay, S., Bera, J. N., Sengupta, S. (2016). Assessment of Saturated Current Transformer Primary Current by Multi Resolution Analysis of Its Secondary Current. *2nd International Conference on Control, Instrumentation, Energy and Communication (CIEC16)*, Kolkata, West Bengal, India. Published in IEEE Xplore Digital Library. https://10.1109/CIEC.2016.7513763.
13. Chattopadhyaya, A., Dey, D., Basu, M., Sengupta, S. (2013, October 10 – 11). Behaviours of Current Transformer under Different Conditions and Its Signal Assessment. *National Conferenceon Recent Development in Electrical, Electronics & Engineering Physics (RDE3P-2013)*, Howrah, West Bengal, India.
14. Chattopadhyaya, A., Banerjee, H., Chattopadhyay, S., Sengupta, S. (2013, October 10 – 11). Effect of Different Switching Transient on CT Saturation. *National Conferenceon Recent Development in Electrical, Electronics & Engineering Physics (RDE3P-2013)*, Howrah, West Bengal, India.
15. Chattopadhyaya, A., Banerjee, H., Chattopadhyay, S., Sengupta, S. (2013, November). Kurtosis Based Assessment of CT saturation Due to Switching Transient. *Second Michael Faraday IET India Summit*, Kolkata, West Bengal, India.
16. Datta, S., Chattopadhyaya, A., Chattopadhyay, S., Das, A. (2020, October, 3-4). *Harmonic Distortion, inter-harmonic Group Magnitude and Discrete Wavelet Transform based Statistical Parameter Estimation for Line to Ground Fault Analysis in Microgrid System*. Kolkata, West Bengal, India.
17. Mukherjee, N., Chattopadhyaya, A., Chattopadhyay, S., Sengupta, S. (2020). Discrete Wavelet Transform and Stockwell Transform based Statistical Parameters Estimation for Fault Analysis in Grid Connected Wind Power System. *IEEE Syst. J.*, 14(3), 4320 – 4328. https://10.1109/JSYST.2020.2984132.
18. Tajdinian, M., Bagheri, A., Allahbakhshi, M., Seifi, A., R. (2018). Framework for current transformer saturation detection and waveform reconstruction. *IET Gener. Transm. Distrib.*, 12(13), 3167-3176. https://10.1049/iet-gtd.2017.1742
19. Odinaev, Pazderin, A., V., Murzin, P., V., Tashchilin, V., A., Samoylenko, V., O., Ghoziev, B. (2021, July 28th – 30th). Detection of the initial region of the current transformer core saturation. *19th International Conference on Renew. Energies and Power Quality (ICREPQ'21)*, Almeria, Spain.

9

Multilevel Inverter-Fed Closed Loop Control and Analysis of Induction Motor Drive

Subrat Behera[1]*, Ranjeeta Patel[1], Rudra Narayan Dash[1] and Amit Kumar[2]

[1]*School of Electrical Engineering, Kalinga Institute of Industrial Technology, Deemed to be University, Bhubaneswar, Odisha, India*

[2]*Thapar Institute of Engineering and Technology Patiala, Deemed to be University, Punjab, India*

Abstract

Multilevel inverters have discovered more extensive extent of utilization in moderate and also in high-power adjustable-speed drives. This chapter introduces a vector control scheme of induction motor drive which includes pulse width modulations for reducing harmonics and total harmonic distortion (THD). For better control of induction motor, indirect vector control has been applied which offers advantages such as removal of flux sensor, more dynamic responses, and minimum torque pulses is applied. The inverter named neutral point clamped inverter is applied for observing dynamic control of the motor drive along with efficiency. The main attention of this chapter is to improve the performance of indirect vector controller. The THD analysis proves the better operation of induction motor as compared to conventional voltage source inverter-fed induction motor drive. By the help of MATLAB simulation, the dynamic performance as well as steady-state of multilevel inverter-based drive are analyzed.

Keywords: Vector control, neutral point clamped inverter (NPCI), synchronously rotating reference frame (SRRF)

**Corresponding author*: subrat.sb@gmail.com

Tamal Roy, Suman Lata Tripathi and Souvik Ganguli (eds.) Linear and Nonlinear System Modeling, (143–158) © 2024 Scrivener Publishing LLC

9.1 Introduction

The acknowledgement engine has just been freed from fixed speed machine to variable speed with the assistance of various control techniques. If a suitable control technique is not implemented properly, the benefits of variable frequency operation cannot be totally recognized. The assumption of method assumes a significant job in deciding the general feasibility, its attributes, and execution of the engine drive. During ordinary activity of the induction drive, the chosen control procedure is needed to guarantee that the activity is restricted to the zone of high working torque. It must be considered that it synchronizes the ratings of inverter which helps to reduce the switching losses as well as the system loss. By the help of control strategies rather than design of the system, we are able to manage the overload conditions or fault conditions. One of the most common motor is induction motor, which is used in most industries, signal control systems, and in the main AC power-driven applications. And the abovementioned are significant benefits of induction motor (IM) drives. It is simple and possesses rugged construction, lower cost, and less maintenance. There are varieties of induction motors available for study. Efficient control of three-phase induction motor requires proper concept of properties of motors. In this work of closed loop induction motor drive, a suitable converter named tandem converter was applied, which proved to minimize the THD of the whole system and display better performance. The induction motor performance encourages the implementation in AC motors. For the better control of induction motor drive several typologies have been developed in the last many years. Decoupling mechanism that works fine in case of separately excited DC motor was applied to AC asynchronous machine. Here, the field oriented control was identified to be suitable for obtaining the outputs. The outputs being the mechanical and electrical as rotor current and speed of the machine. Along with that, the chapter includes the THD performance of the machine using a suitable inverter topology. The approach is to synthesize the machine currents in order to generate better decoupling mechanism. As field oriented control is one of the most suitable approach, the same is used for closed loop control.

The neutral point clamped multilevel inverters (NPCMLI) are also known as diode clamped inverter technically. Here in the voltage level, the middle point is referred to as the inverter's neutral point. The increased inverter's potential has doubled the level of voltage of the device without precise voltage regulation. In various inverter topologies, voltage stress is a central point influencing the operation of multilevel inverters (MLI). In diode clamped MLI, diodes are being implemented in the circuit to restrict the stress on power devices. If N = number of levels, then the voltage sources are counted as [N-1], switching conductor devices are assumed as 2[N-1], and the number of diodes required are [N-1][N-2]. It provides low THD at the output, it operated at both high-level and low-level frequencies, and low rating devices can be used; these are some of the advantages of NPCMLIs [1].

The objective of the chapter is to demonstrate the study and research outcome of NPCMLI of five-level with loads operating at different or wide modulation index. Here, the switches are selected in such a way that it minimizes the common mode voltage (CMV) of the inverter and the influence of the harmonics on the current at the load side is well taken care. After proper analysis of the model and with the result obtained, five-phase, three-level neutral point clamped (NPC) inverter proved dynamic for small changes in load parameters. In this chapter, the study of the model along with the outputs is done with MATLAB simulation [2–4].

Various other researchers worked in quite similar areas which may also be corelated for the future scope of the study. Here the improvement in speed control technique of induction motor drive has offered a wider scope of applications. This study analyzed the reducing harmonic distortions and CMV by using appropriate switching techniques in case of five-level NPC inverter feeding three-phase asynchronous motor drive [5–7]. Mostly, NPC is applied for aircrafts, motor drives, and also in HVDC and AC transmission lines. In this method, transformation of three phases into two phase speed and time dependent system abc to dq is considered. One of the components defines the magnetic flux and other corelates the torque of the motor. Vector control topology is one of the unique concepts which involves decoupling of flux and torque components in order to obtain dynamic responses. Proportional-integral (PI) controllers are typically used in vector control to maintain the measured currents at their reference levels. It improves torque response, dynamic speed accuracy, and low power consumption; these are some advantages of the vector control method. Here, the operation of indirect vector control is discussed in detail.

The fundamental idea behind vector control is to decouple the machine flux and torque which lets us operate on any of the parameters independently without hampering the performance of the other. The instantaneous component of the stator currents here get transformed into a frame rotating at synchronous speed that may be aligned with the stator flux, rotor flux , and airgap flux vectors. In the process, two components, *i.e.*, direct (d)-axis and quadrature(q)-axis components are produced. Both the d and q axes components control the torque as well as machine flux, respectively. The control becomes easier and simpler and the performance of the asynchronous machines becomes dynamic in nature with faster change in load. In our case, the machine used is a three-phase induction motor. The machine drive is considered to be in vector control only if the stator current vector component and the decoupled components in synchronously rotating reference frames (SRRF) match each other.

The makeshift or the transfer of the frames from stationary to rotating is possible when referred to specific flux vectors. Here, the torque component is operated only after the transformation is done. This is otherwise termed as indirect vector control. Here, the challenge or limiting feature of field orientation is being the method that requires detection of the flux angle or technically called flux angle estimation. As per application, the control principle is classified into two categories known as direct and indirect control topology. The flux sensing devices like flux sensing coils or the hall effect sensors are used for performing the task in direct topology.

Field-oriented control uses a framework where the d-axis component is positioned with flux vector of rotor which rotates at the frequency equal to stator. This typical solution enables flux and torque components to be independently operated by the controller system. The rotor flux d-axis component of the stator current is expressed as i_{ds} and the other component of the current is expressed as i_{qs}. The independent nature of the torque and flux is realized only if the position of rotor flux is precisely determined. Generally, with the advantage of the above method, the rotor oriented flux is used more often owing to the fact that alignment of SRRF with the rotor flux is much easier and simpler for obtaining the dynamic behaviors in motor output.

Here, the steady-state and dynamic state capability of the five-level NPC inverter-based drive were analyzed in the MATLAB platform. The control schemes for the neutral point clamped inverter (NPCI) feeding the AC motor drive were illustrated by MATLAB simulations [8–10]. The result analysis was explained along with the THD percentage. In Figure 9.1, the block diagram of three-phase, five-level MLI is shown. Figure 9.2 shows the block diagram of the vector control of an induction motor drive. Figure 9.3 illustrates the simulation of vector control of Induction Motor (IM) drive using neutral point diode clamped inverter in the matlab platform.

Simulation diagram:

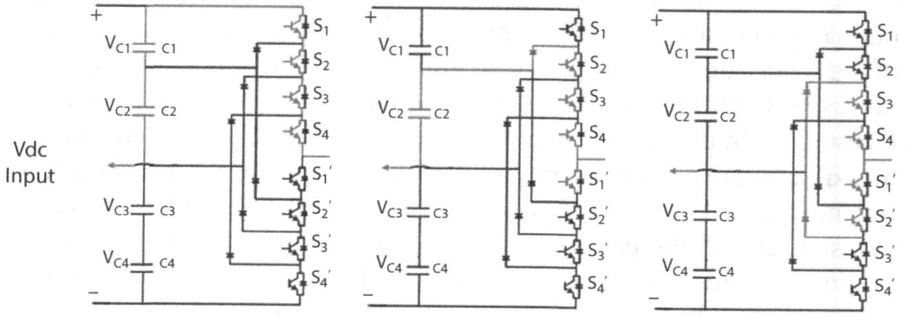

Figure 9.1 Three-phase, five-level NPC inverter.

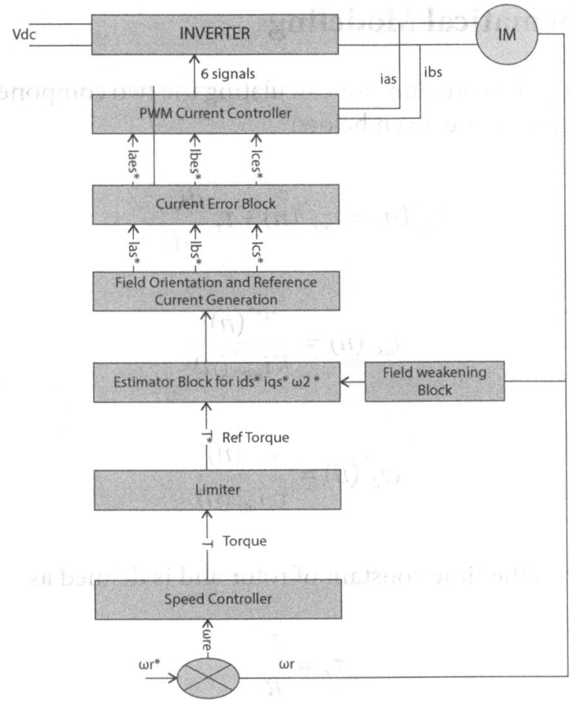

Figure 9.2 Block diagram of closed loop vector control.

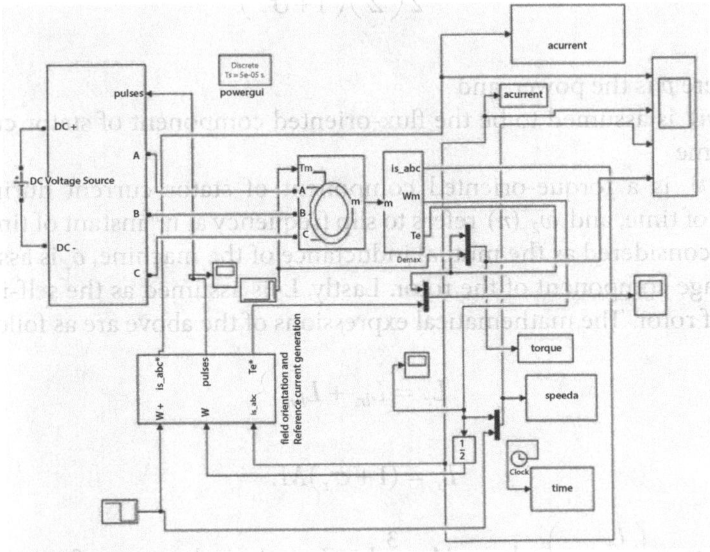

Figure 9.3 Simulation of vector control of IM drive using neutral point diode clamped inverter (NPC).

9.2 Mathematical Modeling

Mathematically, the equations for calculating the two components of d and q axes of the current are given below:

$$i_{ds}{}^{*}(n) = i_{mr}{}^{*}(n) + \tau_r \frac{di_{mr}{}^{*}}{dt} \tag{9.1}$$

$$i_{qs}{}^{*}(n) = \frac{T^{*}(n)}{Ki_{mr}{}^{*}(n)} \tag{9.2}$$

$$\omega_2{}^{*}(n) = \frac{i_{qs}{}^{*}(n)}{\tau_r i_{mr}(n)} \tag{9.3}$$

In which, τ_r is the time constant of rotor and is defined as

$$\tau_r = \frac{L_r}{R_r}$$

$$K = \frac{3}{2}\left(\frac{p}{2}\right)\left(\frac{M.}{1+\sigma_r}\right)$$

Where p is the power, and

$i_{ds}{}^{*}(n)$ is assumed to be the flux-oriented component of stator current at nth time

$i_{qs}{}^{*}(n)$ is a torque-oriented component of stator current during nth instant of time, and $\omega_2{}^{*}(n)$ refers to slip frequency at nth instant of time. $M.$ here is considered as the mutual inductance of the machine, σ_r is assumed as leakage component of the rotor. Lastly, L_r is assumed as the self-inductance of rotor. The mathematical expressions of the above are as follows:

$$L_r = L_{lr.} + L_{lm.} \tag{9.4}$$

$$L_r = (1+\sigma_r)M. \tag{9.5}$$

And $\sigma_r = \left(\frac{L_r}{M} - 1\right)$ where $M = \frac{3}{2}L_m$, L_m is the inductance of magnetism.

Proportional-Integral (PI) Speed Controller:
Here, PI is discussed with the mathematical elaboration. The PI speed controller used here is quite simple and convenient, and is being compared with some other speed controllers.

The torque during nth instant is demonstrated as follows:

$$T_n = T_{n-1} + K_p[\omega_{re(n)} - \omega_{re(n-1)}] + K_i\omega_{re(n)} \tag{9.6}$$

Where T_n is assumed as the torque of the controller at nth time. Gain constants K_p and K_i are referred to as proportionality and integral gains, respectively. $\omega_{re(n-1)}$ acts as net speed at [n−1] instant time [11–13].

The simulation model of the PI speed controller is demonstrated in Figure 9.4(a). As found in the figure, the corresponding and vital increase in parameters, specifically, K_p and K_i, were utilized individually with the limit block and here, the preset value of the torque was determined according to the rating of the motor. In Figure 9.4(b), the modeling was done for estimation of parameters.

9.2.1 Field Weakening Controller

During the weakening of field flux in the closed control of asynchronous motor drive, the prime field weakening system is keenly operated to observe the dynamic behavior of machine output characteristics. The control method is executed at times when the motor speed is measured higher than the minimum speed. The field flux weakening controller system receives input

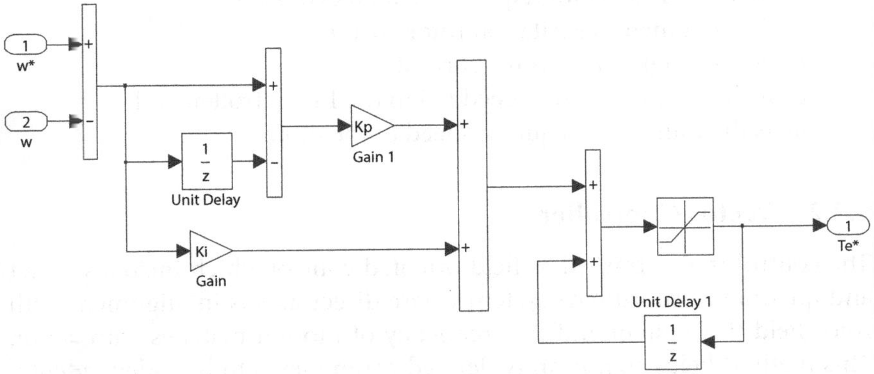

Figure 9.4(a) Model of PI controller.

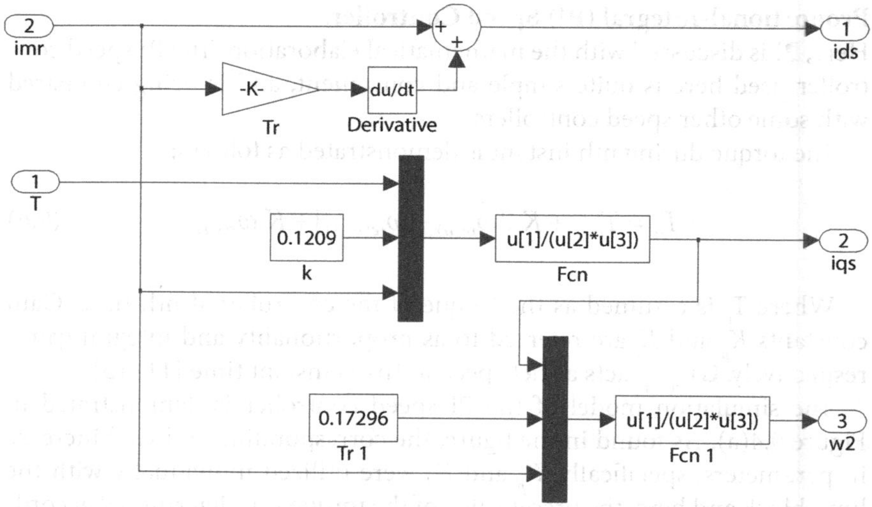

Figure 9.4(b) Parameter estimation model using MATLAB.

regarding the motor's feedback speed. The system's output is the excitation current. The magnetization current remains the same or constant when it is less than the minimum speed. For speed higher the minimum rated speed, the magnetization current changes inversely proportional to the speed [14–19]. The mathematical relation of the mentioned detail is demonstrated,

$i_{mr}{}^* = i_m$, if $\omega_r <$ minimum required value of speed

$i_{mr}{}^* = K_f{}^* \frac{i_m}{\omega_r}$, if $\omega_r >=$ minimum required value of speed. Here,

K_f is the minimum required field flux constant.

$i_{mr}{}^*$ is assumed as excitation rotor current,

i_{mr} is the magnetizing rotor current,

ω_r is the feedback rotor speed fed to the PI controller, and

ω_b is the minimum required speed of the motor.

9.2.2 Vector Controller

The control is also framed as field oriented control which includes direct and quadrature coordinate system where direct-axis is in alignment with rotor field flux system and the frequency of motion matches with stator. This method helps significantly defined components to be independently operated. The current i_{ds} is a d-axis component of stator and caters to the flux. The current i_{qs} is the q-axis component of the stator and is actually controlled for smooth operation. The decoupling or the independence in the operation between these two components is possible only if the identification of the rotor flux is done correctly or measured accurately. Direct flux sensors or flux estimators work fine for the abovementioned [20, 21]. The speed controller produces an output torque and using limiter, (T^*) is considered to be the reference torque. The field weakening controller leads to an output as magnetizing current, which in turn is proportional to flux. These input signals are fed to the control system or field-oriented controller for estimating the components, torque (i_{qs*}), and the flux (i_{ds*}).

Here, the slip frequency is also estimated along with the other two. The basic objective remains to operate the AC asynchronous motor similarly like separately excited DC motor. This phenomenon is achieved as the torque and flux components remain decoupled in the SRRF [22–25].

9.3 Results

Here, the simulation results of the neutral point diode clamped MLIs with resistive-inductive load are presented. The Figures 9.5(a) and 9.5(b) includes the current waveform and harmonic study of five-level MLI inverter simulated in the MATLAB software. The results helped in implementing the inverter feeding the induction motor with specific parameters. The inverter was used for performing closed loop simulation and analyzing the output characteristics. The outputs of the induction motor were also demonstrated in the figures. The outputs can be found as dynamics of speed which are classified as starting dynamics, reversal dynamics, and load perturbations.

Figure 9.5 load current waveform of five-level MLI connected to RL load. The THD analysis is also done on the five-level inverter, as shown in the figure below.

Simulation was performed on a 1-HP induction motor for the control logic and the results are explained for starting and reversal below. The specifications of squirrel cage induction motor are mentioned in the Tables 9.1 and 9.2.

9.3.1 Starting Dynamics

A controlled voltage and frequency source is supplied to the induction motor drive. Here, the value of reference value of speed is fixed at 250 rad/s. The starting torque was limited under the calculated value. The

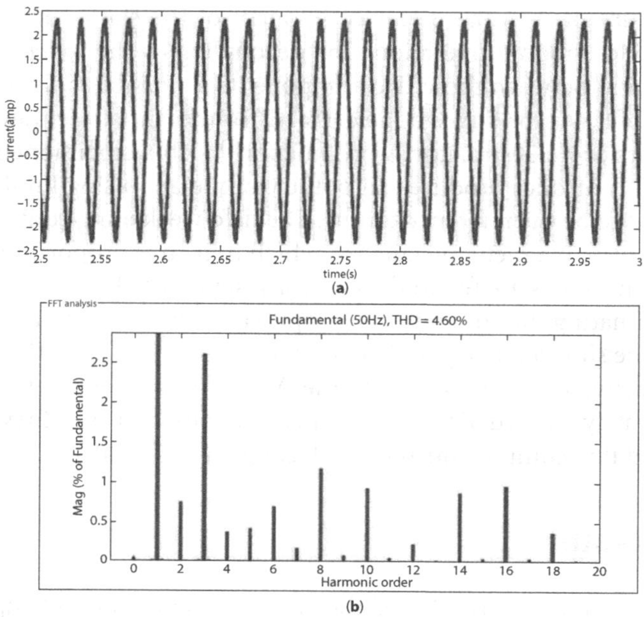

Figure 9.5 (a) Lload current waveform of 5 level MLI connected to RL load. (b) THD analysis of MLI with load current connected to RL load.

Table 9.1 Machine specifications of induction motor used for obtaining results.

Parameters/specifications	Values
Phases	3 AC
Power value in HP	1HP
Voltage value in Volts	415 V
Current in Amps	1.7 A
Pole numbers	4
Frequency used	50 Hz

torque is limited with a value twice the rated value of motor and the result of which the initial current was limited proportional to the value of torque limit while the motor gains the speed at required torque in the beginning. At the speed error reaching nearly equal to zero value in rad/s, the current in the winding decreases to a value equal to no load and the torque developed raises to a value balancing the load torque as found in the responses

Table 9.2 Machine equivalent parameters used for obtaining results.

Equivalent parameters	Value
Stator phase resistance (R_{S})	9.07 Ω
Rotor Phase resistance (R_r)	9.39 Ω
Reactance of stator (X_s)	12.04 Ω
Reactance of Rotor (X_r)	12.04 Ω
X_m/phase (magnetic reactance)	210.999 Ω
Rotational inertia (J)	0.0018Kgm^2

in the figure shown below. The speed reaches the desired value at steady state condition and developed torque becomes minimum, as shown in Figure 9.6.

9.3.2 Reversal Dynamics

In reversal, the speed of the machine was referred to –250 rad/s from 250 rad/s and the controller leads to the development of regenerative braking and phase reversal of the current, which can be visualized in Figure 9.7. The steady state currents are found to be same after reversal and when steady state is reached in the negative direction of motor. Here, when this change in the current signal is observed by the controller, the frequency of

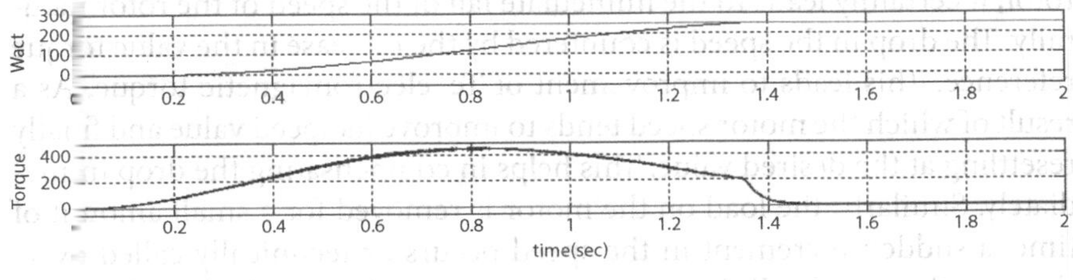

Figure 9.6 Starting dynamics of induction motor.

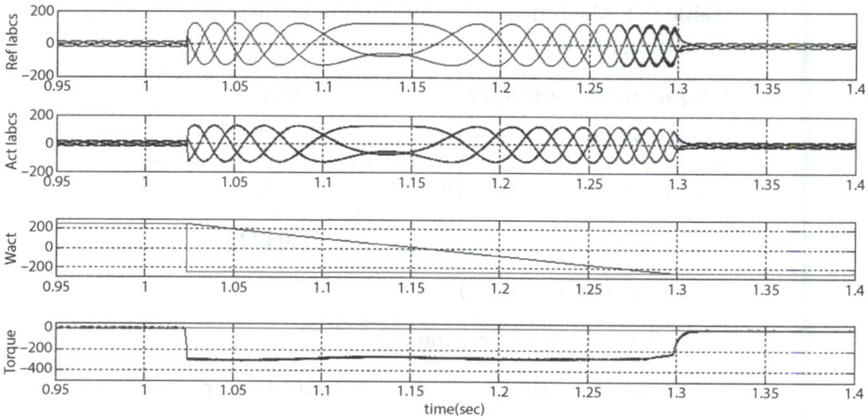

Figure 9.7 Reversal dynamics of induction motor.

the currents in the stator showing the responses of the braking is followed by the reverse of phase current while rotating the rotor in the opposite motion. As the asynchronous motor runs through the dynamic situation at no load before and after the reversal of the rotor rotation, the current of the converter at steady state condition are very much identified to be similar in both directions of the rotor rotation, *i.e.*, starting and reversal of motor rotations. In phase sequences, the current magnitudes are observed to be different in different directions.

9.3.3 Load Perturbation Analysis

The performance analysis of the motor with the above-focused control topology during load perturbation is specifically done as the speed of the drive is tracked and it is not supposed to ideally change under any circumstance of load. The process of study is carried out by implementing and removing the load on the motor drive during its operation at a speed of 250 rad/s at steady state condition. When the load is applied on the rotor, it certainly leads to the immediate fall of the speed of the rotor naturally. The drop in the speed is countered by the increase in the value torque reference. This leads to improvement of Te, electromagnetic torque. As a result of which the motor speed tends to improve its speed value and finally resettling at the desired value. This helps in compensating the drop immediately. Similarly, the load on the motor is removed for a small amount of time, a sudden increment in the speed occurs or technically called overshoot in the speed. All the operations are smoothly handled by the speed controller by keeping the speed fall or rise in the safe limits during constant load variations. High overshoot in speed may lead to instability in maintaining safety in motor application. The motor, as a result, settles at the value of the reference speed. The load perturbations are demonstrated in Figure 9.8. In the figure, it can be observed that the load has been applied and removed at instants and the electrical and mechanical changes in the output characteristics are responding immediately.

The load perturbations are demonstrated in Figure 9.9. The dramatic nature of the load current and speed can be seen in the figure.

9.3.4 THD Analysis

After the analysis, the THD was found to be 15.50%, which is quite appreciable and accepted for driving a motor drive application.

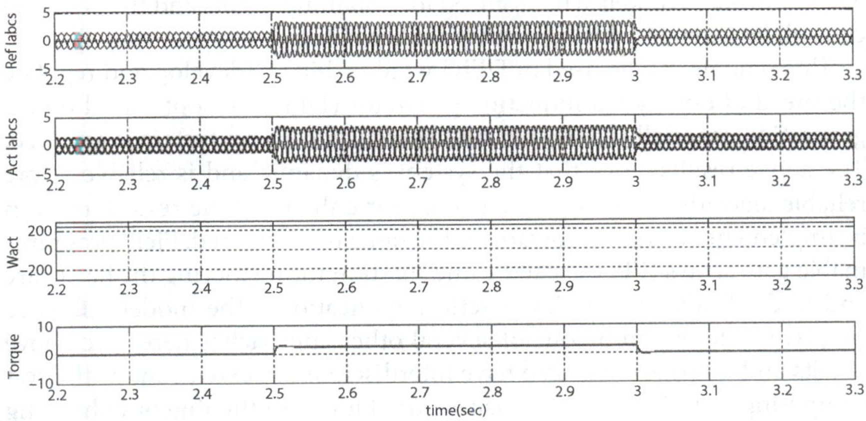

Figure 9.8 Load perturbation of induction motor.

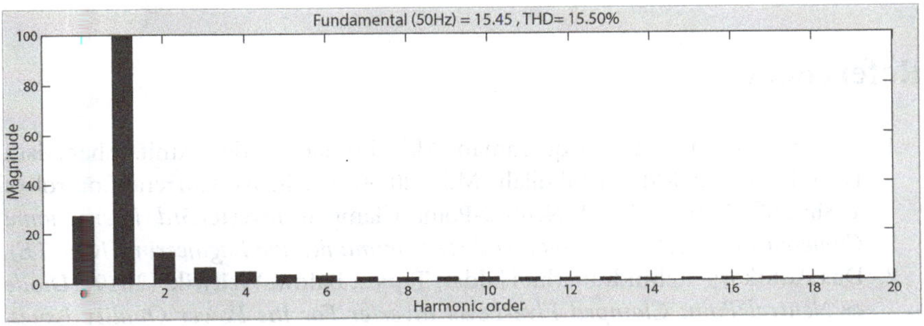

Figure 9.9 THD analysis of rotor current of induction motor.

9.4 Conclusion

The NPC topology was adopted for higher level of power applications as it can achieve better harmonic reduction and the prior existing control strategies help to reduce switching losses. Using this configuration, the stress level of voltage on high frequency power electronics devices is reduced dynamically compared to conventional low level inverters. Initial uses involved the railroad traction areas and steel industries in Japan and Europe. There are several advantages of this structure, which are, major reduction of switching losses in particular and provides low THD at the output. Here, the study was applied to induction motor for the realization of the potential of the methodology. The concept is applicable for smaller and medium applications. But its advantages can be tested on high rated applications. Analysis was done using MATLAB simulation tool and the signals can be generated to realize the same on real time test bed. The converter topology and the loop controller are well synchronized for obtaining results and the potential can still be extended during further studies in future.

This scheme can be used in DFIG wind turbine to develop and regulate the speed of doubly fed induction generator (DFIG) for optimized power generation especially when the wind energy conversion system is grid tied. The above results show that the system is dynamic and is reliable. More reliable algorithm can be implemented for enhancing the results and can be implemented when hardware for the same is considered. Field programmable gate arrays (FPGA) can be one solution for furthering the hardware and find effective results for practical applications. The model reference adaptive systems can be one of several other approaches here. The above results and interface may also have interdisciplinary extensions as the soft computing techniques are extensively used for smoothening or enhancing the applicability. Machine learning algorithms also have a large presence in the recent scenario of technology development.

References

1. Anwer Hossain, Md., Tariquzzaman, Md., Kumar Podder, Amit., Shahadath Hossain Sabbir, Md., Habibullah, Md. (2019). Predictive Current Control of a Simplified Three-Level Neutral-Point Clamped Inverter.*3rd International Conference on Electrical, Computer & Telecommunication Engineering (ICECTE)*.
2. Das Adhikary, Rabindranath., Halder, Tapas., Balow, Writwik. (2019). *Diode & Neutral Point Clamped Five-Level Inverter For the Power Quality Issues.* 978-1-5386-6722-4/19/$31.00 © IEEE.

3. Pillay, T.L., Saha, A.K. (2018). *A comparative analysis of three phase npc multi-level inverter* IEEEPES/IAS.
4. Bheemaiah, Ch., RanjanMuduli, Utkal., Kumar Behera, Ranjan. (2018). Performance Comparison of Five-Phase Three-Level NPC to Five-Phase Two-Level Voltage Source Inverter. 978-1-5386-9316-2/18/$31.00 IEEE.
5. Novotny, D.W. and Lipo, T.A. (1996).*Vector Control and Dynamics of AC Drives*. Oxford University Press.
6. N. Mistry, Jigar., Vakharia, Krishna., Tailor, Dhaval., Patel, Brijesh. (2016). *Software and Hardware Implementation of Five level Neutral Point Clamped Inverter* 978-1-4673-9939-5/16/$31.00 © IEEE.
7. Rodriguez, Jose, Bernet, Steffen, K. Steimer, Peter. (2010) A Survey on Neutral-Point-Clamped Inverters. *IEEE Trans. Ind. Electron.*, VOL. 57, NO. 7, JULY.
8. Rodriguez, J., Bernet, S., Wu, B., Pontt, J.O., and Kouro, S. (2007). Multilevel voltage-source-converter topologies for industrial medium-voltage drives. *IEEE Trans. Ind. Electron.* vol. 54, no. 6, pp. 2930–2945, Dec.
9. Tekwani, P. N., Kanchan, R. S., Gopakumar, K., and Vezzini, A. Fivelevel inverter topology for induction motor drives with common-mode voltage elimination in complete modulation range. (2007). *EPE J*, vol. 17, no. 1, pp. 11-23, January March.
10. IMECS, Maria., Trzynadlowski, Andrzej M. (2005). Vector Control Schemes for Tandem-Converter Fed Induction Motor Drives. *IEEE Trans. Ind. Electron.*, VOL. 20, NO. 2, MARCH.
11. Shiri, A., Vahedi, A., Shoulaie, A. (2006).The effect of parameter variations on the performance of indirect vector controlled induction motor drive.*Power Electronicsand Motion Control 5thInternationalConference* Vol- 2, Page(s):1–5. 14-16 Aug.
12. Texas instruments. (2003). Ac induction motor control (Three phase), system document *Digital control system group*, 30 April.
13. Leite, V., Araujo, R., Freitas, D. (2003) A new online identification methodology for flux and parameters estimation of vector controlled induction motors. *Electric Machines and Drives Conference*, Page(s):449-455 vol. 1. 1-4 June.
14. Barrero, F., Perez, J., Millan, R., Franquelo, L.G. (1999) Self-commissioning for voltage-referenced voltage-fed vector controlled induction motor drives. *Industrial Electronics Society, The 25th Annual Conference of the IEEE*, Page(s):1033-1038 vol.3. 29 Nov. - 3 Dec.
15. Blasco-Gimenez, R., Asher, G.M., Sumner, M., Cilia, J., Bradley, K.J. (1996) Field weakening at high and low speed for sensor less vector controlled induction motor drives.*Power Electronics and Variable Speed Drives, Sixth International Conference*, Page(s):258–261, 23-25 Sept.
16. Krishnan, R., Bharadwaj, A.S. (1991) A review of parameter sensitivity and adaptation in indirect vector controlled induction motor drive systems. *Power Electronics, IEEE Transactions* on Vol-6, Page(s):695–703. Oct.

17. Dr. Tez, E.S. (1995) *A Simple Understanding of Field-orientation for Ac Motor Control*, Loughborough University of Technology, pp: (3/1-3/4),
18. Prof. Zinger, Don S. (1990). *Field Oriented Control of Induction Motors*, University of Akron, Vol-1, pp: 391-396.
19. Krishnan R. (1996). Review of flux-weakening in high performance vector controlled induction motor drives.*Industrial Electronics, Proceedings of the IEEE International Symposium* Page(s):917-922 vol-2, 17-20 June.
20. Ohdachi, Y., Kawase, Y., Hirako, M. (1995) Dynamic analysis of vector controlled induction motor using finite element method. *IEEE Trans.*, Vol-31, Page(s) 1904–1907. May.
21. Singh, Bhim., Singh, B.N., Singh, B.P., A. and Chandra, K., Haddad, Al, (1995) Unity Power Factor Converter-Inverter Fed Vector Controlled Cage Motor Drive Without Mechanical Speed Sensor. *IEEE Trans.*
22. Singh, Bhim., Ghatak, Choudhuri, Sumit. (2005) Simplified Model for Simulation of Vector Controlled Induction Motor Drive (VCIMD) Using Matlab. *International Conference On Computer Applications in Electrical Engineering Recent Advances*, Sep. 29-Oct. 1.
23. Tahami, F., Shojaei. (2006) A Novel Fault Tolerant Reconfigurable Concept for Vector Control of Induction Motors. *Power Electronics and Motion Control, 12th International Conference.* Page(s):1199-1204 Aug.
24. Shiri, A., Vahedi, A., Shoulaie, A. (2006) The effect of parameter variations on the performance of indirect vector controlled induction motor drive. *Power Electronics and Motion Control 5thInternational Conference.* Vol-2, Page(s):1–5. 14-16 Aug.
25. Liu, Xu., Ruan, Yi., Zhang, Chaoyi., Sheng, Huanqing., Yang, Yong. (2008). On Speed Sensorless Vector Control System for Induction Motor Based on Estimating Speed by Torque Current Differential. *Proceedings of the 27th Chinese Control Conference.* July 16-18.

10

Hybrid Grey Wolf Optimizer for Modeling and Control of Electric Drives

Souvik Ganguli[1]* and Prasanta Sarkar[2]

[1]*Department of Electrical and Instrumentation Engineering, Thapar Institute of Engineering and Technology, Patiala, Punjab, India*

[2]*Department of Electrical Engineering, National Institute of Technical Teachers' Training & Research, Kolkata, West Bengal, India*

Abstract

Permanent magnet synchronous motor drives, or PMSM drives for short, use integrated nature-dependent metaheuristic algorithms in this chapter to perform reduced order modeling and controller design in a unified domain. The firefly algorithm (FA) and the grey wolf optimizer (GWO) are two of the most important algorithms in computer science. Together, they form a new hybrid architecture called the hybrid grey wolf optimizer (HGWO). First, a lower-order model of a permanent magnet synchronous motor drive, comprising of speed and current controllers, was obtained through a reduction using a signal-processing-based identification technique. The anonymous proportional-integral (PI) controller gains can be roughly estimated by matching the reduced system developed in cascade with the controller to a desired system. The gains of the delta operator are nearly replicated by the gains of the continuous-time controller. This allows for the development of a unified controller for the drive. Thus, the proposed algorithm is useful not only for order diminution but also for determining the control parameters of PMSM drives. In addition, many commonplace and industrial uses of ac motor drives of this type can benefit in future research from this unified approach study.

Keywords: Firefly algorithm (FA), grey wolf optimizer (GWO), model reduction, controller design, reference model matching

**Corresponding author*: souvik.ganguli@thapar.edu

Tamal Roy, Suman Lata Tripathi and Souvik Ganguli (eds.) Linear and Nonlinear System Modeling, (159–172) © 2024 Scrivener Publishing LLC

10.1 Background Study

Some of the most widely used applications in control engineering produce mathematical relations, typically of higher order. These systems are thus challenging to control. Therefore, it is preferable to reduce its model, as this helps to lessen computational load and issues with higher-order controller implementation. Although some smart approaches to controlling permanent magnet synchronous motor (PMSM) drive have been developed [1, 2], the proportional-integral-derivative (PID) controller continues to enchant industry-related applications thanks to its non-compliant form and merits in a wide range of operations. Some significant literature pertaining to the control schemes of permanent magnet synchronous drives are discussed below, which piques the interest of researchers.

Kumar *et al.* [3] focused on the development and analysis of an electric vehicle (EV) equipped with a PMSM using MATLAB/Simulink. The study investigated the performance of the EV under various drive cycles. The research also aimed to provide insights into the suitability and efficiency of the PMSM-based EV system in real-world driving conditions, contributing to the advancement of electric vehicle technology.

Su *et al.* [4] presented a novel approach for multi-parameter identification of PMSMs using a model reference adaptive system (MRAS) combined with a simulated annealing particle swarm optimization (SA-PSO) algorithm. The study aimed to accurately estimate the key parameters of PMSMs, such as stator resistance, rotor flux linkage, and inertia moment, which are crucial for motor control and performance optimization. The proposed method demonstrated promising results in terms of parameter identification accuracy and robustness, offering potential applications in PMSM control systems.

Zhen *et al.* [5] proposed a Lyapunov-based robust control strategy for the PMSM in the modular joint of a collaborative robot. The research focused on developing a control method that ensures stability and robustness in the presence of uncertainties and disturbances. The proposed approach utilized the Lyapunov stability theory to design a controller that guaranteed tracking accuracy and robust performance of the PMSM. The study also contributed to the advancement of control techniques for PMSMs in collaborative robot applications.

Zhou *et al.* [6] developed an online parameter identification and self-tuning proportional-integral (PI) controller for PMSMs). The study aimed to improve the control performance of PMSMs by continuously updating the motor parameters and adapting the PI controller gains in real-time.

The suggested method enabled accurate parameter estimation and automatic tuning of the controller, leading to enhanced tracking accuracy and robustness of PMSMs in various operating conditions. The research also contributed to the development of intelligent control techniques for PMSMs in robotics and control engineering applications.

Nguyen *et al.* [7] introduced a direct flux-vector oriented control (DFVOC) method for interior permanent magnet synchronous motors (IPMSMs). The article focused on improving the motor's performance and efficiency by accurately controlling the flux and torque in the rotor reference frame. The proposed DFVOC technique simplified the control structure and enhanced the dynamic response of IPMSMs, making it suitable for various applications. The study contributed to the advancement of control strategies for IPMSMs in the field of communications and electronics.

In the study conducted by Huo *et al.* [8], a novel design technique for controlling PMSMs was presented. The researchers proposed a fuzzy adaptive PI regulator that combines fuzzy logic and adaptive algorithms. This approach aimed to enhance the dynamic response and tracking accuracy of PMSMs by adjusting the PI controller gains in real-time based on the motor's operating conditions. The findings of this research contributed to the advancement of control techniques for PMSMs in the field of new energy and electrical technology.

Zhang and Zhu [9] focused on decoupling control techniques for outer rotor coreless bearingless permanent magnet synchronous motors (ORCB-PMSMs). The study investigated the use of LS-SVM (least squares support vector machine) generalized inverse to achieve decoupled control of torque and radial force in ORCB-PMSMs. The work aimed to improve the performance and stability of ORCB-PMSMs by effectively decoupling the control variables. The findings contributed to the development of advanced control strategies for ORCB-PMSMs in the field of electromagnetics.

Zhu *et al.* [10] worked on the development of a nonlinear active disturbance rejection control (NADRC) strategy for PMSM drives. The study explored the application of NADRC to mitigate the effects of disturbances, uncertainties, and nonlinearities in PMSM control systems. The proposed control strategy aimed to improve the dynamic response, robustness, and tracking accuracy of PMSM drives. The research contributed to the advancement of control techniques for PMSMs in the field of energy conversion, offering potential benefits in terms of performance and efficiency.

Khorashadizadeh *et al.* [11] presented an adaptive control mechanism of robot manipulators driven by PMSMs using the orthogonal functions theorem. The study investigated the application of this theorem in designing adaptive control algorithms for PMSM-based robot manipulators.

The proposed technique aimed to enhance the tracking performance and robustness of the manipulator system by adapting control parameters based on the system dynamics. Further, the research contributed to the advancement of control techniques for PMSM-driven robot manipulators, offering potential benefits in terms of precision and stability.

In their work, Zhong *et al.* [12] proposed a fractional-order feedforward control method for permanent magnet linear synchronous motors (PMLSMs) based on the frequency-domain adjustment theory. The study aimed to enhance the dynamic performance and tracking accuracy of PMLSMs by utilizing the fractional-order control approach. By incorporating frequency-domain adjustments, the method proposed allowed for improved control of PMLSMs, enabling more precise and efficient operation. The investigation thus contributed to the advancement of control techniques for PMLSMs in the field of mechanical systems and signal processing.

In their study, Chen *et al.* [13] suggested a simplified model predictive control (MPC) method for a twelve-phase PMSM. The work aimed to improve the control performance of the PMSM by reducing computational complexity while maintaining accurate control. The proposed simplified MPC approach offered a practical solution for real-time control of the twelve-phase PMSM, contributing to the advancement of control techniques in the field of industrial electronics.

Kumar and Singh [14] proposed a modified model reference adaptive control method for sensorless control of a PMSM-driven light electric vehicle (LEV). The study also focused on developing an online self-tuning speed controller to improve the accuracy and efficiency of the PMSM control system. The proposed approach offered a practical solution for sensorless control of PMSM-driven LEVs, contributing to advancements in power electronics, drives, and energy systems.

It is generally accepted that continuous sampling is the foundation upon which discrete systems are built. Modeling, assessing, and regulating these systems all make use of a method called z-transform analysis. Having their poles so close together near (1, 0) makes it impossible for these systems to remain stable at a high sampling rate. The delta operator, developed by Middleton and Goodwin [15], can be used to build such discrete-time systems. Thus, the systems must be simulated in the delta domain so that high-sampling-rate continuous-time systems can be constructed and better control over shift-operator-based discrete-time systems in terms of computation can be improved. Therefore, delta operator-based systems are being used more frequently in research and design of systems [16–18].

Among synthesis-based methods used in control engineering, the Truxal technique is highly prevalent [19]. The controller parameters are determined by the degree to which the entire closed-loop system with controller in cascade conforms to the reference model, which is achieved by first adjusting the reference model to meet certain time and frequency domain standards. The primary shortcoming of the exact matching model [20] is the absence of a working controller. One possible answer is to use approximate model matching (AMM) on the unified domain of analysis [21]. Ganguli *et al.* [22] developed the unified domain to reduce higher-order systems with a hybrid grey wolf optimizer (HGWO). This algorithm has been expanded to use the AMM method for tuning controller parameters in the delta domain and minimizing the higher-order PMSM drive model.

The remaining text of this chapter is written in the style shown. Section 10.2 explains the methodology used in the work. The simulation results are described and discussed in Section 10.3. Section 10.4 discusses the final thoughts on the matter.

10.2 Proposed Approach

This chapter focuses on two major themes. Ganguli *et al.'s* [22] hybrid technique is used in the first part to reduce the order of the PMSM drive model. The second involves creating a PI controller with the same heuristic approach. Polynomials in the numerator and denominator are found using the HGWO during the model reduction process. The reduced model with a fixed structure is implemented using the following three constraints.

- Harmonizing the original's steady-state gain with the predicted low-order system
- The reduced system must have poles with negative real parts that are equivalent to those of the original system.
- Keeping the original model's non-existence of zeroes on the right half of the s-plane.

When assessing the appropriateness of a model's settings, the sum of squares of errors (SSE) is often used. The model identification procedure takes as input a pseudo random binary sequence (PRBS).

Hybrid grey wolf optimizer (HGWO), a combination of the firefly algorithm (FA) technique [23] and the grey wolf optimization (GWO) method [24], is used to reduce the number of higher-order single-input-single-output (SISO) systems in the delta domain. The FA is used to make global adjustments in the exploration phase, while the grey wolf optimization methods is utilized as a local search mechanism to verify exploitation capabilities. The method is activated with the aid of FA to kick off the quest. The calculation is repeated for a fixed number of times to find the optimal global position in the necessary solution space. Therefore, the GWO algorithm begins with the result that FA generated. Then, GWO is used as the search method to accelerate the convergence to the optimal outcome. The combined solution of FA and GWO methods can successfully find an optimal value in this fashion.

The particle swarm optimization (PSO) [25], the Harris hawks optimization (HHO) [26], the arithmetic optimization algorithm (AOA) [27], the war strategy algorithm (WSA) [28], and the wild horse optimization (WSO) [29] are some of the methods used to compare and contrast with the proposed method. For the sake of comparison, we also use a method that combines the chaotic grey wolf approach with PSO. The prime motive behind model order reduction approach is to carry out the control synthesis for a practical low order PI controller in this chapter. The PID parameters are typically obtained through auto-tuning or by using traditional old PID tuning rules [30]. In the Truxal approach [19], which is used in precise model matching control [20], the coefficients of the reference system are calculated to achieve the desired outcome, and the controller gains are assessed in a way that is very similar to the complete closed-loop system. The most significant drawback of this exact model matching (EMM) method is that the controller's implementation in hardware is not assured. To get around this limitation of the EMM procedure, AMM [21] will be used instead of the current design schemes. Therefore, the HGWO technique is used in conjunction with AMM technology to retrieve the controller's coefficients. Similarly, a set of alternative algorithms is used to compare with the HGWO method.

10.3 Simulation Outcomes and Discussions

The input-output relation of the PMSM drive system [31, 32] is represented with the help of

$$G_{HOS}(s) = \frac{7.94s + 13.23}{9.6\times10^{-7}s^3 + 3.99\times10^{-3}s^2 + 6.97s + 34.63} \tag{10.1}$$

Ultimately, we want to create a PMSM drive controller that works well. Due to the higher-order nature of the input-output connection produced from the available data, the first step is to simplify the model so that it can be used with a normal PI controller. Due to its ability to reduce drive-related noise and interruption, PI controllers are frequently chosen over traditional PID controllers. For a sampling rate of about 1 kHz, the discrete-delta domain transfer function of the aforementioned model is denoted by Equation 10.1 as

$$G_{HOS\delta}(\gamma) = \frac{100\gamma^2 + 1200300\gamma + 1995900}{\gamma^3 + 2040\gamma^2 + 1061200\gamma + 5.224200} \tag{10.2}$$

When creating the reduced models in the delta domain, we assume population sizes of 20 and 100 and use a maximum of 100 iterations. All published algorithms are considered, with their default parameters (proposed, parent, and standard heuristics) taken into account. The unified domain's reduced systems are listed in Table 10.1. The minimum values for the fitness function are also listed for your convenience. The optimal outcome is also highlighted in italics in Table 10.1. The ancestor procedures and a few other methods are used to draw comparisons.

Some well-known strategies in the literature have been compared with the proposed HGWO method. This approach, as demonstrated in Table 10.1, has the lowest fitness value. Table 10.2 details the estimated transient parameters for the simplified models. Results in the table with bolded results are the most similar to the model's parameters. In addition to the methods used for comparing, the newly developed chaotic grey wolf optimizer-particle swarm optimization (CGWOPSO) techniques are also utilized in this table. Ten standard one-dimensional chaotic maps [33] are used to develop CGWOPSO-1 to CGWOPSO-10 algorithms.

Once again, the integrated method outperforms the alternative algorithms discussed in this section. In addition, the reduced-order models are evaluated using the standard error metrics from systems and control theory. The integral of absolute error (IAE) is calculated. It is first of all compared with the ten variants of the hybrid technique CGWOPSO. Figures 10.1 and 10.2, respectively, show the comparison of HGWO with the different chaotic versions used with the hybrid CGWOPSO methods in two parts.

From Figures 10.1 and 10.2, it is clear that the IAE value for the proposed HGWO method is quite less than as compared to CGWOPSO methods. Further, the parent techniques as well as standard methods are also compared with HGWO in terms of IAE. The graph is shown in Figure 10.3.

The error corresponding to IAE with the suggested HGWO method is not only less than the parent FA and GWO approaches but also found less than some of the latest methods with which comparison is carried out

Table 10.1 Model reduction and the minimum cost function of the test system in the unified domain of analysis.

Methods	Second-order models in the unified domain	Minimum cost function
HGWO (proposed)	$\frac{990.5\gamma + 1436.0144}{\gamma^2 + 843.3\gamma + 3759.2}$	***2.9563e–04***
FA [23]	$\frac{989.4\gamma + 2435.2118}{\gamma^2 + 843.8\gamma + 6374.9}$	3.5926e–04
GWO [24]	$\frac{990.5\gamma + 3685.918}{\gamma^2 + 849.5\gamma + 9649}$	3.0213e–04
PSO [25]	$\frac{989.7\gamma + 3742.4158}{\gamma^2 + 845.4\gamma + 9796.9}$	3.9686e–04
HHO [26]	$\frac{990.4767\gamma + 155.4423}{\gamma^2 + 841.9306\gamma + 406.9172}$	4.2561e–04
AOA [27]	$\frac{993\gamma + 3800.6708}{\gamma^2 + 851.5\gamma + 9949.4}$	3.4352e–04
WSA [28]	$\frac{954.2497\gamma + 1995.6}{\gamma^2 + 836.4\gamma + 5224.2}$	5.2345e–04
WHO [29]	$\frac{991.6\gamma + 3786.193}{\gamma^2 + 788.44\gamma + 9911.5}$	3.0635e–04

in Figure 10.3. Other metrics of performance like ITAE, ISE, and ITSE can also be tried out in future to provide the assessment of the proposed method over the techniques available in the literature.

The final step now is to use the integrated delta domain to determine the PI controller's settings. For controller synthesis, the AMM method is employed. The proposed algorithm allows for a heuristic determination of the controller's parameters; Table 10.3 displays the obtained results. Along with determining the optimal values for the regulatory parameters, the heuristic scheme also determines the respective fitness value as well. Once again, bold letters denote the bare minimum fitness level.

Table 10.2 Transient specifications of the second-order PMSM drive model in the unified domain.

PMSM model	Methods	Rise time (t_r)	Settling time (t_s)	% peak overshoot
Test system (original)		0.0027	0.7803	204.4378
Lower-order models	HGWO	**0.003**	**0.8909**	**202.8970**
	CGWOPSO-1	0.0034	0.6286	201.4254
	CGWOPSO-2	0.004	0.5549	200.4539
	CGWOPSO-3	0.004	0.5693	199.4859
	CGWOPSO-4	0.0035	0.6288	201.9586
	CGWOPSO-5	0.0037	0.6123	201.3424
	CGWOPSO-6	0.004	0.5798	198.4981
	CGWOPSO-7	0.004	0.5343	200.0289
	CGWOPSO-8	0.004	0.5107	200.1456
	CGWOPSO-9	0.0036	0.6103	201.0902
	CGWOPSO-10	0.0038	0.5924	201.4562
	FA [23]	0.004	0.5205	200.1621
	GWO [24]	0.004	0.5367	200.2378
	PSO [25]	0.004	0.3397	197.2367
	HHO [26]	0.004	0.5578	200.9283
	AOA [27]	0.004	0.5483	201.9921
	WSA [28]	0.004	0.5547	201.2548
	WSO [29]	0.004	0.5778	200.1459

The proposed hybrid algorithm obtains the minimum optimum value when compared to some existing heuristic techniques. Due to its ability to efficiently generate lower-order PMSM models and make use of approximate model matching technology, the proposed method is able to generate a controller (proportional+integral) with the minimum cost function when compared to the original and other alternate methods.

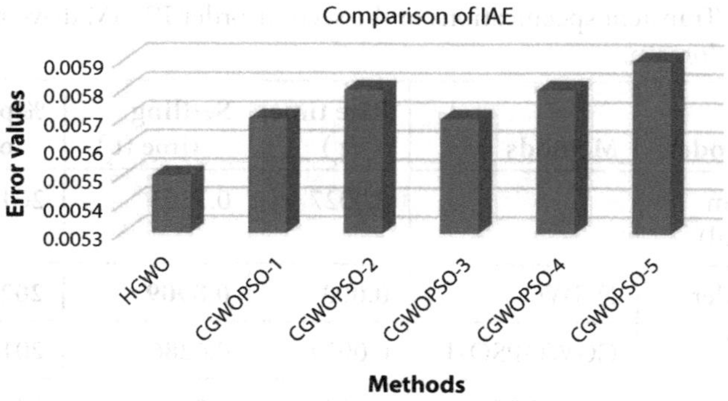

Figure 10.1 Comparison of IAE between HGWO and CGWOPSO 1–5 methods.

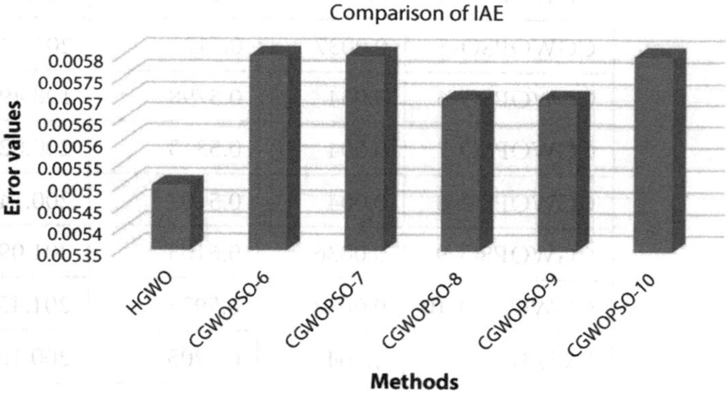

Figure 10.2 Comparison of IAE between HGWO and CGWOPSO 6–10 methods.

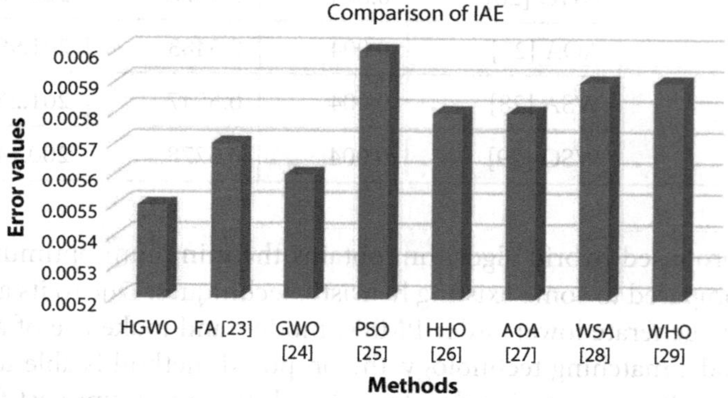

Figure 10.3 Comparison of IAE between HGWO and some standard heuristics.

Table 10.3 Parameters of the controller and the cost function values of the reduced-order PMSM drive system in the combined domain.

Algorithms	K_P	K_I	Fitness function value
HGWO	1.8679	1.4967e-05	**7.3715e-05**
GWOPSO-1	0.0010	4.3879	7.6759e-05
GWOPSO-2	0.0041	2.1390	7.9023e-05
GWOPSO-3	0.0187	2.8646	7.3853e-05
GWOPSO-4	0.0009	4.3969	7.6776e-06
GWOPSO-5	0.0011	3.2887	7.7516e-05
GWOPSO-6	0.0019	4.3868	7.7428e-05
GWOPSO-7	0.0011	4.3923	7.6763e-05
GWOPSO-8	0.0027	4.3908	7.8690e-05
GWOPSO-9	0.0010	4.3888	7.6759e-05
GWOPSO-10	0.0014	4.3861	7.6850e-05
FA [23]	0.7172	3.3084e-04	7.5926e-05
GWO [24]	1.8012	1.3171e-04	7.388e-05
PSO [25]	0.8281	4.4387	7.7782e-05
HHO [26]	0.0013	4.3772	7.6833e-05
AOA [27]	0.0043	2.2451	7.5692e-05
WSA [28]	0.0025	4.4372	7.6717e-05
WSO [29]	0.0021	4.3328	7.5622e-05

10.4 Conclusions

The PMSM drive's discrete delta domain controller architecture and order reduction are discussed here. The PMSM drive's reduced-order models are enhanced using a constrained global optimization method. For this purpose, a hybrid approach involving the GWO and FA methods was used effectively. In addition, its PI controller's parameters are derived using ideas from approximate model matching. This technique is advantageous because it relies solely on the feedback from the output. The control mechanism was developed in such a way that it closely tracks the target model. In this chapter, we also discussed one practical use of the combined domain approach for modeling and control of the PMSM drive system. Similarly, induction motor drives, which are a common type of drive, can have their size minimized and a controller designed for them using the same kinds of smart methods. Similar modeling and control methods can be applied to other common drives, like switched reluctance motor (SRM) and brushless direct current (BLDC) motor drive systems.

References

1. Patil Rahul, D., Saravana Prakash, P., Sandeep, N., & Verma, A. K. (2022). Performance Evaluation of PMSM Drive Using ANFIS based Speed Controller Under SVPWM Technique. In *2022 IEEE 10th Power India International Conference (PIICON)* (pp. 1-6).
2. Fatemimoghadam, A., Yan, Y., Iyer, L. V., & Kar, N. C. (2022). Permanent Magnet Synchronous Motor Drive Using Deep-Neural-Network-Based Vector Control for Electric Vehicle Applications. In *2022 International Conference on Electrical Machines (ICEM)* (pp. 2358-2364).
3. Kumar, A., Chandekar, A., Deshmukh, P. W., & Ugale, R. T. (2023). Development of electric vehicle with permanent magnet synchronous motor and its analysis with drive cycles in MATLAB/Simulink. *Mater. Today Proc.*, 72, 643-651.
4. Su, G., Wang, P., Guo, Y., Cheng, G., Wang, S., & Zhao, D. (2022). Multiparameter identification of permanent magnet synchronous motor based on model reference adaptive system—Simulated annealing particle swarm optimization algorithm. *Electronics*, 11(1), 159.
5. Zhen, S., Li, Y., Liu, X., Wang, J., Chen, F., & Chen, X. (2023). A Lyapunov-based robust control for permanent magnet synchronous motor in the modular joint of collaborative robot. *Robotica*, 41(5), 1389-1406.
6. Zhou, W., Hua, Q., Yao, Y., Kong, L., Xie, A., & Zhang, D. (2022). On-line parameter identification and self-tuning PI controller for permanent magnet synchronous motor. In *2022 2nd International Conference on Robotics and Control Engineering* (pp. 26-31).
7. Nguyen, T. P., Hsieh, M. F., Huynh, T. A., & Doan, Q. V. (2022). Direct Flux-Vector Oriented Control for Interior Permanent Magnet Synchronous Motor. In *2022 IEEE Ninth International Conference on Communications and Electronics (ICCE)* (pp. 501-506).
8. Huo, D., Yang, Z., Wang, Y., Wang, B., & Gong, C. (2023, March). Novel Design Technique of Fuzzy Adaptive PI Regulator for Permanent Magnet Synchronous Motor. In *Proceedings of the 3rd International Symposium on*

New Energy and Electrical Technology (pp. 657-669). Singapore: Springer Nature Singapore.

9. Zhang, Z., & Zhu, H. (2022). Decoupling control on outer rotor coreless bearingless permanent magnet synchronous motor using LS-SVM generalized inverse. *Prog. Electromagn. Res. M*, 111, 65-76.
10. Zhu, L., Zhang, G., Jing, R., Bi, G., Xiang, R., Wang, G., & Xu, D. (2022). Nonlinear active disturbance rejection control strategy for permanent magnet synchronous motor drives. *IEEE Trans. Energy Convers.*, 37(3), 2119-2129.
11. Khorashadizadeh, S., Zirkohi, M. M., Eliasi, H., & Gholipour, R. (2022). Adaptive control of robot manipulators driven by permanent magnet synchronous motors using orthogonal functions theorem. *J. Vib. Control*, 10775463221085784.
12. Zhong, Y., Gao, J., & Zhang, L. (2023). Fractional-order feedforward control method for permanent magnet linear synchronous motor based on frequency-domain adjustment theory. *Mech. Syst. Signal Process.*, 190, 110115.
13. Chen, B., Lv, J., & Jiang, X. (2019). Simplified Model Predictive Control of a Twelve-Phase Permanent Magnet Synchronous Motor. In *IECON 2019-45th Annual Conference of the IEEE Industrial Electronics Society* (Vol. 1, pp. 1345-1350).
14. Kumar, S., & Singh, B. (2022). Modified Model Reference Adaptive Sensorless Control of PMSM Driven LEV with Online Self-Tuning Speed Controller. In *2022 IEEE International Conference on Power Electronics, Drives and Energy Systems (PEDES)* (pp. 1-6).
15. Middleton, R. H., & Goodwin, G. C. (1990). *Digital Control and Estimation: A Unified Approach (Prentice Hall Information and System Sciences Series)*. Englewood Cliffs, NJ: Prentice Hall.
16. Huang, C., Naghdy, F., & Du, H. (2017). Delta operator-based fault estimation and fault-tolerant model predictive control for steer-by-wire systems. *IEEE Trans. Control Syst. Technol.*, 26(5), 1810-1817.
17. Kumari, K., Bandyopadhyay, B., Kim, K. S., & Shim, H. (2019). Output feedback based event-triggered sliding mode control for delta operator systems. *Automatica*, 103, 1-10.
18. Ramírez-Neria, M., Luviano-Juárez, A., Lozada-Castillo, N., Ochoa-Ortega, G., & Madonski, R. (2020). Discrete-Time Active Disturbance Rejection Control: A Delta Operator Approach. In *Advanced, Contemporary Control* (pp. 1383-1395). Springer, Cham.
19. Truxal, G. J. (1955). *Automatic Feedback Control System Design*. McGraw-Hill, New York.
20. Tzafestas, S. G., & Pimenides, T. G. (1982). Exact model-matching control of three-dimensional systems using state and output feedback. *Int. J. Syst. Sci.*, 13(11), 1171-1187.
21. Wang, S.S., & Chen, B.S. (1987). Model matching: optimal approximation approach. *Int. J. Control*, 45(6), 1899-1907.

22. Ganguli, S., Kaur, G., & Sarkar, P. (2020). A new hybrid algorithm for identification in the unified delta framework. In *AIP Conference Proceedings* (Vol. 2207, No. 1, p. 040002). AIP Publishing LLC.
23. Yang, X. S. (2010). Firefly algorithm, stochastic test functions and design optimisation. *Int. J. Bio-Inspired Comput.*, 2(2), 78-84.
24. Mirjalili, S., Mirjalili, S. M., & Lewis, A. (2014). Grey wolf optimizer. *Adv. Eng. Software*, 69, 46-61.
25. Kennedy, J., & Eberhart, R. (1995, November). Particle swarm optimization. In *Proceedings of ICNN'95-International Conference on Neural Networks* (Vol. 4, pp. 1942-1948).
26. Heidari, A. A., Mirjalili, S., Faris, H., Aljarah, I., Mafarja, M., & Chen, H. (2019). Harris hawks optimization: Algorithm and applications. *Future Gener. Comput. Syst.*, 97, 849-872.
27. Abualigah, L., Diabat, A., Mirjalili, S., Abd Elaziz, M., & Gandomi, A. H. (2021). The arithmetic optimization algorithm. *Comput. Methods Appl. Mech. Eng.*, 376, 113609.
28. Ayyarao, T. S., RamaKrishna, N. S. S., Elavarasan, R. M., Polumahanthi, N., Rambabu, M., Saini, G., ... & Alatas, B. (2022). War strategy optimization algorithm: a new effective metaheuristic algorithm for global optimization. *IEEE Access*, 10, 25073-25105.
29. Naruei, I., & Keynia, F. (2021). Wild horse optimizer: A new meta-heuristic algorithm for solving engineering optimization problems. *Eng. Comput.*, 1-32.
30. O'Dwyer, A. (2009). *Handbook of PI and PID controller tuning rules*. Imperial college press.
31. Ayyar, K. (2013). *Investigations on model order reduction and design of speed controller for electric drive systems.*
32. Krishnan, R. (2001). *Electric motor drives: modeling, analysis, and control* (Vol. 626). New Jersey: Prentice Hall.

11

Parameter Estimation of First-Order RC Model of Lithium-Ion Batteries in Electric Vehicles Using Slime Mould Algorithm

Ramdutt Arya[1], Shatrughan Modi[2] and Souvik Ganguli[3*]

[1]*Centre for Energy & Environment, Malaviya National Institute of Technology, Jaipur, Rajasthan, India*
[2]*School of Computing, Indian Institute of Information Technology, Una, Himachal Pradesh, India*
[3]*Department of Electrical and Instrumentation Engineering, Thapar Institute of Engineering and Technology, Patiala, Punjab, India*

Abstract
Electric vehicles (EVs) are being considered as the promising solution for carbon-free transportation and the future of mobility. Battery electric vehicles (BEV) are the true EVs that completely run from electricity supplied by their batteries (Li-ion). For dynamic simulation of BEV and use of the battery effectively, it is essential to model the battery and estimate its parameters accurately. Generally, estimation of the battery parameters requires complex, time-consuming, and expensive methods. This chapter focuses on a simple equivalent circuit methods (ECM) is used for battery modeling. Estimation of battery parameters is done by comparing the proposed model output to the known catalogue output with the help of a newly formed heuristic optimizer-slime mould algorithm (SMA). Slime mould algorithm is based on the morphological transformations, oscillation, and foraging of slime mould found in nature, and have shown remarkable exploration and exploitation capabilities. The evaluation of 31 parameters for first-order ECM is done by taking the fitness function as the sum of squared-error of model voltage and catalogue voltage, for charging as well as discharging scenarios. Additionally, four well-known algorithms, ant lion optimizer (ALO), dragonfly algorithm (DA), grey wolf optimizer (GWO), and sine-cosine algorithm (SCA), are examined for comparison of performance and efficiency in terms of the parameters, accuracy

**Corresponding author*: souvik.ganguli@thapar.edu

Tamal Roy, Suman Lata Tripathi and Souvik Ganguli (eds.) Linear and Nonlinear System Modeling, (173–195)

(minimized error), and convergence. A statistical study of the mean, standard deviation, and best and worst values for each method is also performed to ensure that the results are stable and reliable. Finally, a Wilcoxon rank-sum test with Holm-Bonferroni adjustments and a Kruskal-Wallis test are used to evaluate and verify the significance of the data. For first-order ECM, performance is slightly better than other algorithms.

Keywords: First-order RC model, parameter estimation, slime mould algorithm (SMA), Wilcoxon test, Kruskal-Wallis test

11.1 Introduction

The global environment is facing huge challenges. Air pollution and GHG (Green House Gases) emissions due to the burning of fossil fuels are among the major challenges. This burning of fossil fuel is done mainly in the transportation, power, and industrial sectors. The transport sector obtains 90% of its energy need from burning of fossil fuels [1]. Invalid source specified. In order to decrease dependency on fossil fuels and replacing them with clean and environmentally friendly solutions, the electrification of the transport sector is the main focus of every nation. In this perspective, electric vehicles (EVs) have become the best choice to fulfil the need for a clean source of transportation that produces very low or no carbon emission and give enhanced fuel economy. This has increased the demand for EVs as these are now being considered as the favorable transportation mode [2] Invalid source specified.

Electric vehicles are those vehicles in which there is an electric drive or vehicles in which propulsion power (either fully or partially) is received from the electricity. There are mainly three types of EVs: hybrid electric vehicle (HEV), plug-in hybrid electric vehicle (PHEV), and battery electric vehicle (BEV). Among these, BEVs are considered as the true EV and fully green mode of transport if the battery is charged from the green sources only [3].

Electric vehicles' success mainly depends on their battery/battery pack. The battery serves as the energy source for the EV. This is a crucial part of the EV, as most of the important aspects related to EV like range, charging time, cost, weight, size, and safety are mainly dependent on it [4]. The battery pack consists of various cells packed and connected in a specific manner to achieve a certain voltage/power requirement. These cells are made of materials that possess electrochemical properties. There are various types of batteries available in the market based on materials like lead-acid batteries, lithium-ion (Li-ion) batteries, nickel-cadmium batteries, and nickel-metal hydride batteries [5]. Lithium-ion batteries are the most favorable type of batteries to use in an EV because of their highly specific power, low self-discharge rate, high energy density, high efficiency, and long operating life [6]. The battery mentioned further in this chapter will refer to the Li-ion battery unless otherwise specified.

There are various parameters associated with the battery like terminal voltage, open circuit voltage (OCV), battery internal resistance, temperature, cycle life, charging/discharging rate, state-of-charge (SoC), and depth-of-discharge (DoD). Some parameters are directly measured while others are estimated through various techniques. To facilitate this and provide an interface between the designer/user and the battery, every EV is installed with a BMS (battery management system). The BMS ensures the reliable, efficient, safe, long, and cost-effective usage of the battery by accurately estimating and evaluating all associated parameters. For this, a BMS has to be advanced enough and equipped with a precise parameter estimation algorithm [7].

The battery behavior and characteristics are nonlinear and complex due to its internal electrochemical properties. Capturing exact behavior and accurate battery parameter estimation through various battery modeling techniques are still under study. Several battery models have been developed with different capabilities and different levels of accuracy and complexity. Broadly, these models are categorized in four types [8], *viz.*, electrochemical model—based on detailed electrochemistry of battery, experimental—based on rigorous experimentation, mathematical—based on stochastic approach, and electric circuit model (ECM)—based on electrical equivalent or lumped model.

Plett [9] deliberated upon few equivalent circuits or say lumped models like zero-hysteresis, one-state hysteresis, simple, and enhanced self-correcting (ESC) combined models. Another set of equivalent circuit model which is based on resistance-capacitance (RC) networks such as first-order RC equivalent model [10], second-order RC equivalent model Invalid source specified.[11], and third-order RC equivalent model [12], which represent dynamic characteristics of the battery. These RC models were further compared for feasibility, complexity, and accuracy, especially for Li-ion batteries. It was observed that as the order of the RC model increases, the accuracy, complexity, uncertainty, and cost increase [13]. Battery models needed for real-time applications like in EV must be simple, fast, and accurate. So, in this chapter, we will be using the first-order RC equivalent circuit model for battery parameter estimation.

From various studies, it is observed that the precise estimation of the internal battery parameters is a difficult task because of its nonlinear behavior. Traditional methodologies for battery parameter estimation use electrochemical impedance spectroscopy (EIS), rigorous electrochemical modeling, etc. [8]. These methodologies are complex, time-consuming, expensive, and require special equipment and facilities. Also, many a time, it is required to use different batteries from different ratings and manufacturers to assess and study the performance of a particular EV. It is not feasible and efficient to perform all the detailed methods of estimating the battery parameters. Also, conventional algorithm-based methods are quite complex and require high computation power, which includes the chance of some numeric uncertainties.

In order to resolve the abovementioned issues, the authors in Reference [14] suggested battery parameter estimation as a problem solvable by optimization techniques and used a multi-objective genetic algorithm-based approach for estimation. In Reference [15], authors used the genetic algorithm (GA), grey wolf optimizer (GWO), and particle swarm optimization (PSO) for battery parameter estimation and compared the performances. Analysis showed that the accuracy of the GWO is best as compared to other metaheuristic optimization techniques used in battery parameter estimation. Metaheuristic optimization techniques are simple, flexible, use derivation-free mechanisms, and are superior in local-optima avoidance. Due to these advantages, these metaheuristic techniques have emerged as the best choice for solving optimization problems [16]. Although various metaheuristic optimization and soft-computing techniques have been developed to estimate the parameters of the batteries, there is not yet a metaheuristic technique best suited for finding solutions to all optimization problems, which is proved by mathematical logic in the no free lunch (NFL) theorem [17].

In this chapter, for estimation of battery parameters, we used a newly formed metaheuristic optimization technique named slime mould algorithm (SMA) [18], which has shown remarkable exploration and exploitation capabilities in various optimization problems. Detailed statistical analysis and non-parametric assessment are also performed to validate the efficiency, reliability, and significance of the results obtained from SMA with four other well-known metaheuristic algorithms, *viz.*, ant lion optimizer (ALO), dragonfly algorithm (DA), GWO, and sine-cosine algorithm (SCA). This chapter is further organized as per the following sections. Section 11.2 consists of mathematical modeling of the first-order RC equivalent model of battery. Section 11.3 explains the SMA, Section 11.4 explains the methodology followed. Section 11.5 covers the simulation results and discussion. Concluding remarks with future scope of work is delineated in Section 11.6.

11.2 Brief Overview of the Battery Models

The battery is considered the most important component of an EV. There are different types of batteries available in the market based on the materials used for the electrodes, based on the electrolyte used, etc. Among all these types, Li-ion batteries are most suitable for EVs due to their unique characteristics like large energy density, large specific power, long cycle life, less self-discharge rate, low maintenance, and safe use [6].

For optimum and safe utilization of the battery, continuous monitoring and necessary steps in real-time are required. To achieve this, various parameters associated with batteries like SoC, charging rate (Cr), discharging rate (Dr), and DoD are estimated with the help of various battery models. These models capture the complex and dynamic battery circuit features. For utilization in the EV, these models are supposed to be simple and precise. In literature, these models are categorized mainly in four categories:

1. Electrochemical models - these models are highly precise, however, necessitates detailed knowledge of the electrochemistry of the batteries and use complex nonlinear differential equations [8].
2. Experimental models - these models involve various types of experimental analysis to evaluate the internal parameters of the batteries. These models are not convenient and best fit for depicting the cell dynamics in order the estimate the SoC of battery packs [19].
3. Mathematical models - these models follow the heuristic approaches to anticipate the runtime behavior of battery packs. These models and the previous two models, generally, are quite complex and require intensive computation. Due to this, it is difficult to use these models in circuit simulations to anticipate the performance. Also, these models are not well-suited for real-time power management [19].
4. Equivalent circuit models (ECM) – ECM [10] have high potential with respect to parametrization and accuracy in representing the dynamic characteristics of EV batteries. These are lumped models and use very few numbers of

parameters. This is the simplest equivalent circuit model of the battery which consists of a voltage source and a resistor in series with it. This series resistor denotes the battery's internal resistance. This simple battery model does not consider the dynamic behavior of the battery. It also does not account for SoC, Cr, and Dr. In Reference [8], the author discussed a few other equivalent circuit models or say lumped models like zero-hysteresis, one-state hysteresis, simple, and ESC combined models. These models were further compared with the test data sets of an HEV Li-polymer battery cell [13].

Another set of equivalent circuit models which are based on resistance-capacitance networks such as first-order RC ECM, second-order RC ECM, and third-order RC ECM, represent dynamic characteristics of the battery. These models use a parallel RC circuit connected in series with the internal resistance and the voltage source. The single parallel RC pair is known as the Warburg impedance. The number of these RC circuits determines the order and the complexity of the model [11, 20].

11.2.1 Equivalent Circuit Model (ECM) of Li-Ion Battery

Equivalent circuit models (ECMs) for Li-ion batteries are quite simple and based on Thevenin's equivalent. These ECMs can predict the current-voltage (I-V) characteristics of the battery accurately. To construct an ECM with RC circuit, we use open-circuit voltage (V_0) with a series resistance (R_0) as equivalent ohmic-resistance and one parallel combination of resistance-capacitance pair, C_1 and R_1, as shown in Figure 11.1 [20]. From the electrochemical point of view, current collector and terminal resistances are represented by the series resistance whereas charge-transfer phenomena, double-layer capacitance, and reaction kinetics are represented by the RC pair. The parameters of the battery are represented by the multivariable functions of SoC, DoD, C_r, D_r, cycle number, and temperature number, which have nonlinear behavior [15].

Higher-order models where $n \geq 2$ are complex and require more computational power than the first-order RC model. Complexity, computational power, cost, and unpredictability generally increase as the order increases [12].

11.2.2 First-Order RC Equivalent Circuit Model

The first RC model is the simplest model in the RC ECM group. The parallel RC network in which R_1-C_1 characterizes the self-discharge energy and simulates the transient response whereas the parameters V_0 represents the usable capacity, and R_0 characterizes the instantaneous voltage-drop [15].

These battery parameters depend on multi-variable functions of SOC_{cr}, DOD_{cr}, C_r, D_r, cycle number, and temperature, which have nonlinear behavior. These parameters have an exponential nature [20] and are represented as polynomials functions, formulated in the below equations to express the electrical behavior of the battery.

$$R_0 = (\alpha_1 + \alpha_2 x + \alpha_3 x^2)e^{-\alpha_4 y} + (\alpha_5 + \alpha_6 x + \alpha_7 x^2) \tag{11.1}$$

$$R_1 = (\alpha_8 + \alpha_9 x + \alpha_{10} x^2)e^{-\alpha_{11} y} + (\alpha_{12} + \alpha_{13} x + \alpha_{14} x^2) \tag{11.2}$$

$$C_1 = -(\alpha_{15} + \alpha_{16} x + \alpha_{17} x^2)e^{-\alpha_{18} y} + (\alpha_{19} + \alpha_{20} x + \alpha_{21} x^2) \tag{11.3}$$

$$\begin{aligned} V_0 &= (\alpha_{22} + \alpha_{23} x + \alpha_{24} x^2)e^{-\alpha_{25} y} \\ &\quad + (\alpha_{26} + \alpha_{27} y + \alpha_{28} y^2 + \alpha_{29} y^3) - \alpha_{30} x + \alpha_{31} x^2 \end{aligned} \tag{11.4}$$

where α_1, α_2, α_3, ……, α_{31} are coefficients of the equations formulated as a mix of polynomials and exponential and V_0 represents the terminal voltage when no-load is connected (open circuit voltage).

Equations 11.1–11.5 are used to estimate the parameters and terminal voltage of the battery when the load is connected to its terminals for charging as well as discharging scenario for the first-order RC equivalent model. During operations when current is kept constant, the time-dependent, nonlinear expression for the battery terminal voltage is represented by the following polynomial equation [14],

$$V^M = V_0 + \left(\frac{Q_0}{C_1} + IR_1\right)e^{-t/R_1 C_1} - I(R_0 + R_1) \tag{11.5}$$

Here, $Q_0(0)$ is the nominal rated capacity of the battery at the time of start and 't' is the time duration since charging and discharging starts, both these values are provided by the manufacturer.

In Equations 11.1–11.4, by replacing x with C_r and y with SoC_{cr} for charging scenario, and replacing with D_r and $(1\text{-}DoD_{dr})$ for discharging scenario, respectively, we can obtain the parameters for the battery model. For discharging scenario, $(1\text{-}DoD_{dr})$ is chosen, as with the increase in DoD_{dr}, V_0 decreases [15].

Charge rate (C_r) and D_r can be controlled as per requirements of the application. The C_r and D_r are kept constant for a specific BEV operation and can be determined from the following equations,

$$Cr = \frac{I}{Q_0} \tag{11.6}$$

$$Dr = \frac{I}{Q_0} \tag{11.7}$$

The SOC_{cr} and DOD_{dr} can be determined from the following equations,

$$SoC_{cr} = SoC_{ini} + \left(\frac{I\Delta t}{Q_0.3600} \right) \tag{11.8}$$

$$DoD_{dr} = DoD_{ini} + \left(\frac{I\Delta t}{Q_0.3600} \right) \tag{11.9}$$

Using the above equations, the model which is formed can be deployed as a battery pack in any BEV simulation after estimation of polynomial coefficients and battery parameters along with the battery charging and discharging characteristics identification [15].

11.2.3 Fitness Function for Optimization

Optimization is mainly used in estimating the battery model parameters for first-order RC equivalent model using Equations 11.1–11.4. These are used to obtain the voltage-SoC characteristic curve for this model, which best fits the voltage-SoC characteristics curve taken from the manufacturer's battery.

Here, the fitness is formulated in the form of sum of the squared error between voltage values obtained from the model presented and catalogue values at various sample data points for both chargings as well as discharging cases. This fitness function can be mathematically expressed as

$$F(x) = \sum_{i=1}^{n} (V_i^M - V_i^C)^2 \tag{11.10}$$

with constraint as

$$0 < \alpha_1, 0 < \alpha_2, 0 < \alpha_3, \ldots .0 < \alpha_{31}$$

where V_i^C denotes the ith voltage value taken from the catalogue and V_i^M is the ith value of model output voltage obtained from Equation 11.5 for the first-order RC equivalent. The optimization is done with the help of SMAa newly formed metaheuristic optimization technique at different SoC and DoD values.

11.3 Slime Mould Algorithm (SMA)

Slime mould algorithm (SMA) is a newly formed metaheuristic optimization algorithm that is developed based on the diffusion, oscillation mode, morphological changes, and foraging of the slime mould found in nature. With its exceptional exploration capacity and exploitation intensity, slime mould uses the bio-oscillator-based idea of adaptive weights to plot out the most efficient route to its food source [18].

Slime mould is formed by an organism composed of one or more cells with visibly evident nuclei and organelles. It is found in cool, moist, and dark spots. Plasmodium is the primary nutritional phase and it is also the dynamic and active stage of slime mould. In this stage, organic content explores food, slowly wraps the food from all sides, and releases enzymes so that it can absorb it. Further, during the process of migration, the mould's front end expands into a fan-shaped structure. After this is a network of veins that carry cytoplasm throughout the body. They are able to construct the venous structure connecting the meals because they have distinctive patterns and properties that allow them to absorb nutrients from multiple sources simultaneously [21].

When a vein gets close to a food supply, the bio-oscillator sends out a wave that spreads outward. This enhances the flow of cytoplasm inside the vein. This rate of flow of cytoplasm determines the thickness of the vein, *i.e.*, fast flow will lead to a thick vein and vice versa. With the help of this positive and negative feedback combination, the optimal path can be established by the slime to connect the food source in better and efficient way [22].

It is a three-stage process to discover an optimal path to the food. The first one is searching and approaching the food that depicts the exploration, oscillation, and wrapping of the food, which demonstrates the exploitation characteristics of slime mould [18].

➢ **Approaching the food**

According to the intensity of the smell of the food in the air, the slime mould approaches the food. Mathematical formula for this approaching behavior is given as below [18]

$$X(t+1) = \begin{cases} X_b(t) + vb.\left(W.X_A(t) - X_B(t)\right), & r < p \\ vc.X(t), & r \geq p \end{cases} \quad (11.11)$$

where vb is a parameter that varies from $-a$ to a and the parameter vb follows the linear variation from one to zero. t denotes the current iteration, X_b is the position of an individual from the population where the odor density found is highest, X denotes the position of slime mould, X_A and X_B are the two individuals selected randomly from the population of slime mould, W denotes the adaptive weight of the slime-mould, and p is represented by the below equation [18]

$$p = tanh|S(i) - DF| \quad (11.12)$$

here, $i \in 1,2,3,\ldots,n, S(i)$ denotes the fitness of all individuals from X, DF denotes the best-fitness attained so far in all iterations.

The equation of vb is as below

$$vb = [-a, a] \quad (11.13)$$

where the formula for a can be given

$$a = arctanh\left(-\left(\frac{t}{max\_t}\right)+1\right) \tag{11.14}$$

and the formula for W is given as below [29]

$$W(Smellindex(i)) = \begin{cases} 1+r \cdot log\left(\frac{bF-S(i)}{bF-wF}+1\right), & condition \\ 1-r \cdot log\left(\frac{bF-S(i)}{bF-wF}+1\right), & other \end{cases} \tag{11.15}$$

$$SmellIndex = sort(S) \tag{11.16}$$

here, *condition* tells that half of the population is ranked by $S(i)$, r represents the random value in the [0,1] interval. max_t shows the maximum number of iterations, bF and wF represents the best fitness and the worst fitness value, respectively, obtained in the current iteration. *SmellIndex* represents the sorted fitness values.

➤ **Wrapping the food**

This process demonstrates the stage of contraction of the venous tissue structure of slime mould while searching for food. The bio-oscillator generates a stronger wave whenever the vein contacts a higher concentration of food. Due to this, the cytoplasm flows faster which results in a thicker vein. The positive and negative feedback between the width of the vein and the food density that was explored is mathematically simulated by Equation 11.15 [18].

➤ **Oscillation**

The propagation wave, generated by the biological oscillator, redirects the slime mould's cytoplasm in its veins to a more favorable location in terms of food density. Variations of the venous width of slime mould are being simulated by W, vb, and vc. Synergistic characteristics between vb, and vc resemble the selective nature of slime mould. Once it finds a good source of food, it keeps trying to search for a better source of food. To achieve a higher quality of food from the previous one, it separates some of its organic matter to explore other locations, instead of investing all of it only a single source of food [18].

11.4 Methodology

In this work, SMA is employed. This metaheuristic algorithm is used to estimate the 31 polynomial coefficients of the first-order RC equivalent circuit model of the battery, for the charging scenario as well as discharging scenario. It further helps to determine the modeled battery parameters $R_{0,}$ R_1, C_1, and V_0, which are nonlinear multivariable functions of SOC_{cr}, DOD_{cr}, C_r, and D_r. The values of C_r and D_r are taken as 1 and remains constant for a particular BEV. The values of SOC_{cr} and DOD_{cr} were taken from the manufacturer datasheet. These modeled battery parameters are estimated with four other well-known metaheuristic algorithms, *viz.*, ALO, GWO, DA, and SCA using Equations 11.1–11.4, which in turn calculate the terminal voltage from Equation 11.5 and compared with catalogue values to determine the best fitness using the fitness function, Equation 11.19. For the discharging scenario, we performed the abovementioned process except that we used the value of x as D_r and y as (1-DOD_{cr}) because the value of terminal voltage is decreasing with the increase in discharge.

The optimal parameters are estimated from the SMA and the other four standard algorithms taken as reference. The SMA-estimated optimal parameters are then evaluated with those obtained from alternative optimization methods. The rate of convergence in estimating parameters and fitness values from each algorithm is also determined. The fitness value is defined as the sum of squared error. For the statistical analysis, we also determined the best and worst fitness values, as well as the mean and standard deviation of these values.

Apart from the detailed statistical analysis, the obtained results are were tested for significance. To check the significance, there are various parametric and non-parametric tests. Parametric tests are used when the data is uniformly distributed or follow some pattern, while non-parametric tests are those in which there is no uniformity in the data or the data is minimal [23]. In our case, the obtained results data do not have any pattern.

We used the non-parametric Wilcoxon rank-sum test to obtain the results of our first-order ECM for charging as well as discharging case. This test provides the p-values for every algorithm compared with the proposed algorithm [24]. Generally, the p-values, which are less than 0.05, confirm that the results obtained by the proposed algorithm are significant in comparison with other algorithms [25]. This p-value can be decided by our requirements to check the significance. Here, we have taken it as a standard choice, *i.e.*, $p < 0.05$.

The Holm-Bonferroni correction method is applied after getting the p-values from the Wilcoxon rank-sum test. This correction method adjusts the p-values in such a way that any ambiguity or error that occurred during the calculation of p-values is mitigated [26].

We also performed another non-parametric test named Kruskal-Wallis which gives a graphical representation of the significance of the obtained results. It is a popular technique for the comparison of more than two samples [27].

11.5 Simulation Results and Discussions

For the purpose of the estimation of the parameters, we used an energy innovation group (EIG) Li-ion battery having 8 Ah as the rated capacity and nominal voltage of 2.5 V. The battery specifications are given in Reference [28]. The voltage-SoC curve characteristics during charging and discharging of the battery for various charging rates and discharging rates are from the catalogue.

Slime mould algorithm (SMA) is utilized for parameter estimation of the first-order RC model. The parameters are also estimated with four other metaheuristic algorithms as stated above for comparison purposes. The population size and the maximum number of iterations for each algorithm are taken as 30 and 500, respectively. The dimension size is 31 for this first-order RC ECM. Each algorithm is being run 20 times. The optimal parameters, best and worst fitness values, along with mean and standard deviation, are studied carefully. The optimal parameter values are reported in Tables 11.1 and 11.2.

The obtained values of optimal parameters with SMA are quite close to the optimum values of the parameters from the literature. From the convergence point of view, for the first-order charging scenario, there is not much difference in the convergence of SMA from other algorithms. Still, the SMA is performing a little bit better than other algorithms, as we can see in Figure 11.1(i). While in the case of the first-order discharging scenario, it can be seen from Figure 11.1(ii) that the SMA is converging faster than the DA and a little bit faster than GWO and ALO. The convergence of the SMA and SCA are quite similar.

From a statistical point of view, the fitness value (sum of squared error) analysis for each algorithm is tabulated in Table 11.3. In this table, the best, worst, and mean fitness values, along with the standard deviation, are determined for analysis purposes. For the charging scenario, the SMA is performing well except for the ALO. However, for the discharging

Table 11.1 Optimal parameters α_1 to α_{31} of first-order RC equivalent model-charging scenario.

	Charging				
	SMA	DA	GWO	ALO	SCA
α_1	0.19026	1.42E−01	0.03142	0.2592	0.08407
α_2	0.00325	0.08461	0.14588	0.16657	0
α_3	0.00149	4.93E−02	0.01149	0.08442	0
α_4	35.5184	38.3434	41.5743	32.1945	37.4525
α_5	0.11365	1.22E−01	0.07509	0.10695	0
α_6	0.01709	1.75E−02	0	3.20E−02	0
α_7	0.00099	2.00E−03	5.29E−04	0.00478	0
α_8	0.10669	0.16532	0.06618	0.29677	0
α_9	0.00042	0.01284	0.00377	0.03043	0
α_{10}	0.12277	2.93E−02	0.07064	0.04574	9.87E−02
α_{11}	23.315	30.2619	31.4038	26.3019	27.6545
α_{12}	0.13582	0.14868	0.15231	0.1524	0.0656
α_{13}	0.00784	8.40E−03	1.35E−03	0.0084	0
α_{14}	0.06243	8.88E−02	0.03006	0.07917	0.07973
α_{15}	207.049	232.598	272.081	208.619	263.866
α_{16}	1.13579	0.01767	0.46359	1.18202	0
α_{17}	0.43917	0.16892	0.33342	1.17935	0
α_{18}	16.6422	24	17.5118	16.1644	22.8696
α_{19}	699.366	704.04	693.278	602.102	509.972
α_{20}	1.17513	0.25752	1.03647	1.13982	0.01756
α_{21}	0.80647	0.00823	1.02686	1.11308	0
α_{22}	0.888	0.01842	0.43112	0.66406	0.75837

(Continued)

Table 11.1 Optimal parameters α_1 to α_{31} of first-order RC equivalent model-charging scenario. (*Continued*)

	Charging				
	SMA	**DA**	**GWO**	**ALO**	**SCA**
α_{23}	0.85209	0.858	0.03277	0.85203	0.37747
α_{24}	0.6024	0.46379	0.50208	0.6024	0
α_{25}	16.288	18.696	16.288	16.288	17.3324
α_{26}	0.09074	0.18304	0.11127	0.264	0
α_{27}	0.01553	0.0156	0.0156	0.0156	0.01268
α_{28}	0.00031	1.20E−03	1.83E−04	0.0012	0.00E+00
α_{29}	0.00022	0.00E+00	0.22564	0.234	1.24E−01
α_{30}	0.00633	0.678	0.12319	0.65862	0
α_{31}	0.00096	1.45E−02	0.00146	0.0324	0

scenario, the SMA is outperforming the other algorithms, which can be observed from Table 11.3. This signifies the reliability and the efficiency of the SMA over other algorithms.

Apart from the statistical analysis of the obtained results, we performed the non-parametric assessment to check and validate the significance of the obtained results. The Wilcoxon rank-sum test, Kruskal-Wallis test, and rank-sum test with Holm-Bonferroni corrections were used for this purpose.

The test results for the Wilcoxon rank-sum test are tabulated in Table 11.4. We can see that the *p*-values for the SMA, when compared with other metaheuristic algorithms, are less than 0.05, *i.e.*, the results are significant and null-hypothesis is rejected [29, 30].

The adjusted *p*-values of the proposed algorithm for the Wilcoxon test after applying Holm-Bonferroni corrections are mentioned in Table 11.5. From this table, we again see that the *p*-values are much less than the threshold value of 0.05. This also gives the validation of the significance of the results.

Table 11.2 Optimal parameters α_1 to α_{31} of first-order RC equivalent model-discharging scenario.

Discharging					
	SMA	**DA**	**GWO**	**ALO**	**SCA**
α_1	0.00358	3.60E−03	0.00229	0.002	2.63E−03
α_2	0.2544	0.2544	0.2544	0.2544	0.2544
α_3	0.0316	0.0316	0.0313	0.026	3.45E−03
α_4	28.013	28.3173	27.992	27.992	27.992
α_5	0.43786	0.4428	0.44263	0.43183	2.63E−01
α_6	0.00077	9.92E−02	1.08E−01	0.19197	2.12E−01
α_7	0.27464	4.10E−01	0.15582	0.35246	2.86E−01
α_8	5.26E−16	0.00793	1.19E−04	0.00018	0.00E+00
α_9	2.72E−11	3.60E−03	2.45E−05	0.00194	0.00178
α_{10}	1.69E−13	8.31E−05	1.34E−03	0	0.00E+00
α_{11}	33.3559	39.552	27.241	26.368	36.8966
α_{12}	3.27E−08	1.43E−05	0.0004	2.36E−03	0.00E+00
α_{13}	2.06E−05	6.97E−04	9.83E−04	0.00119	0.00E+00
α_{14}	5.77E−06	0.01318	8.57E−05	0.00404	4.14E−03
α_{15}	131.073	102.936	133.586	104.083	102.208
α_{16}	0.00057	1.1616	0.57839	0.52199	0
α_{17}	0.02776	0.44499	0.009	1.02224	0
α_{18}	20.3842	20.712	14.5013	13.8092	18.2589
α_{19}	718.283	544.919	547.835	516.275	515.12
α_{20}	0	1.27E−01	0.01687	0.20793	2.24E−01
α_{21}	0.03202	1.48E−01	0.08251	0.09718	5.16E−02
α_{22}	0.04258	0.02997	0.02201	2.53E−02	0.00E+00
α_{23}	0.00512	4.32E−03	0	0.0096	0.0095

(*Continued*)

Table 11.2 Optimal parameters α_1 to α_{31} of first-order RC equivalent model-discharging scenario. (*Continued*)

	Discharging				
	SMA	**DA**	**GWO**	**ALO**	**SCA**
α_{24}	0.06051	0.02562	0.01094	0.05337	3.75E−02
α_{25}	29.908	27.2212	28.3482	24.1168	28.1856
α_{26}	4.25335	5.35951	3.87943	4.41369	4.26034
α_{27}	6.61E−13	0.05437	0.04162	0.02355	0
α_{28}	0.12911	1.09E−01	0.11337	0.11593	1.39E−01
α_{29}	0.0624	2.98E−02	6.92E−04	0.06222	6.24E−02
α_{30}	0.31674	3.17E−01	0.04519	0.30802	0
α_{31}	3.89115	4.45181	3.89873	5.69485	3.86E+00

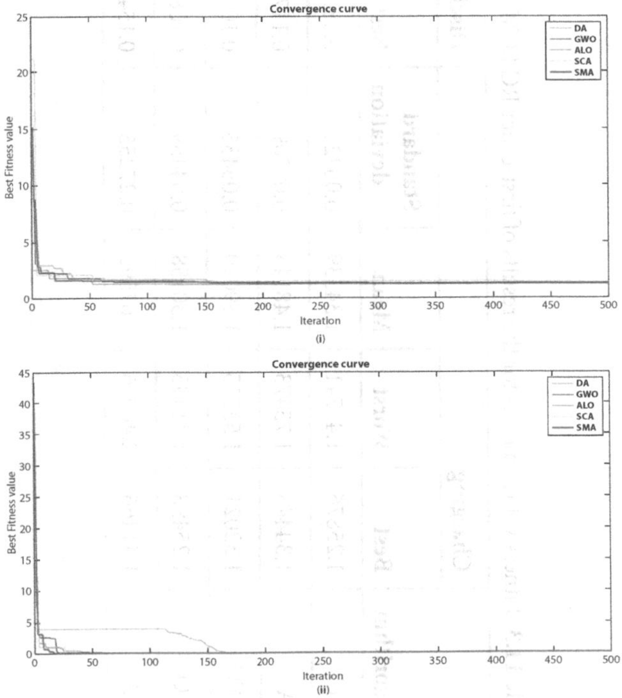

Figure 11.1 (i) Convergence curve for the first-order ECM charging. (ii) Convergence curve for the first-order ECM discharging.

Table 11.3 Fitness value analysis for the results of first-order RC ECM.

Algorithm	**Charging**				**Discharging**			
	Best	**Worst**	**Mean**	**Standard deviation**	**Best**	**Worst**	**Mean**	**Standard deviation**
SMA	1.25876	1.47741	1.34339	0.06224	0.08789	0.13355	0.10736	0.01183
DA	1.34443	1.75473	1.48143	0.09209	0.11789	2.57200	0.88574	0.65209
GWO	1.32021	1.53535	1.39020	0.05435	0.10643	0.19519	0.14297	0.02716
ALO	1.25498	1.41495	1.30158	0.04459	0.10831	0.79773	0.17523	0.14888
SCA	1.41466	2.40946	1.67595	0.27255	0.15419	6.62060	1.24497	1.67679

Table 11.4 Wilcoxon rank-sum test for first-order RC ECM (*p*-values).

Battery ECM	Proposed method	*p*-values			
		ALO	**DA**	**GWO**	**SCA**
First order charging	SMA	2.5639E-02	5.1658E-06	1.6669E-02	1.6571E-07
First order discharging	SMA	3.4995E-06	1.2346E-07	2.3025E-05	6.7956E-08

Table 11.5 Modified *p*-values for Wilcoxon test for first-order RC ECM with addition of Holm-Bonferroni corrections.

Battery ECM	Proposed method	Corrected *p* - values			
		ALO	**DA**	**GWO**	**SCA**
First-order charging	SMA	2.5639E-02	1.5497E-05	3.3338E-02	6.6283E-07
First-order discharging	SMA	6.9989E-06	3.7039E-07	2.3025E-05	2.7182E-07

The Kruskal-Wallis test provides us with the graphical representation of the mean ranks for the first-order ECM. Figure 11.2(i) shows the mean ranks for the charging scenario. In this figure, the proposed SMA group (Group 1) of results are shown with a blue color line while other groups (Groups 2–4) are shown with red color lines if these are significantly different from Group 1. If the other groups' results are not significantly different, it is shown in grey color (faded out). Here, we can see that the mean rank Group 1 is different from the mean ranks of Groups 3 and 5, whereas the mean rank of Group 1 is not different from the mean ranks of Groups 2 and 4, and we can see these are overlapping to some extent.

In Figure 11.2(ii), we can see that the mean rank of Group 1 is different from all other groups, *i.e.*, Groups 2–5. None of the groups is overlapping with Group 1.

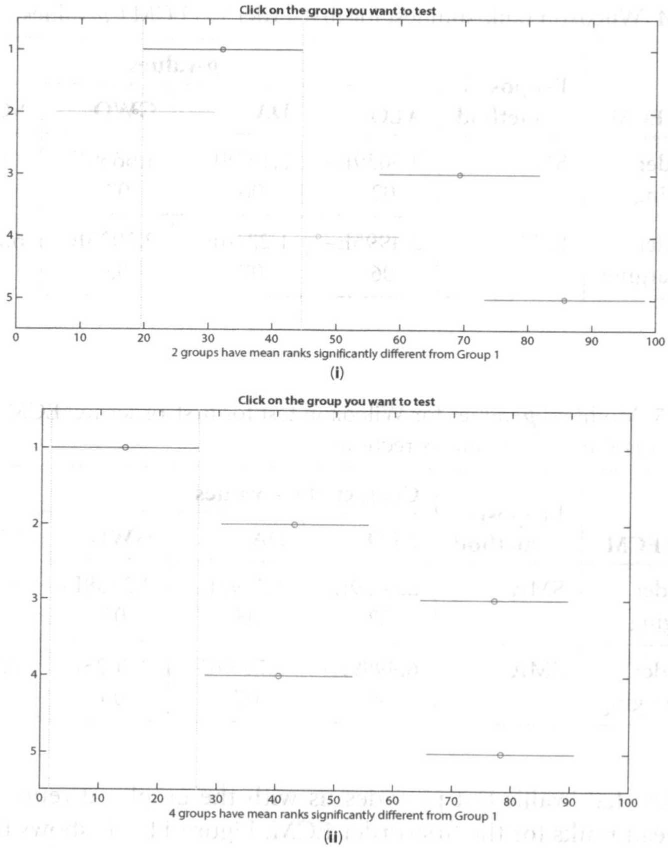

Figure 11.2 (i) Kruskal-Wallis test results for the First-order ECM charging. (ii) Kruskal-Wallis test results for the first-order ECM discharging.

11.6 Conclusions

This chapter demonstrates the mathematical modeling of the first-order RC ECM of battery which constitutes 31 parameters for charging as well as discharging scenarios. The battery chosen is of Li-ion type, which is the most suitable for EV applications. Then, the estimation of the parameters of this model is done with the help of a newly formed metaheuristic technique named SMA. The results obtained from SMA are then contrasted with four other standard metaheuristics techniques like DA, GWC, ALO, and SCA. Simple statistical analysis like mean, standard deviation, and best and worst fitness values are done. The compelling and statistical results demonstrate that SMA has the advantage over other algorithms in the sense that the sum of square error is the least as compared to standard algorithms. Slime mould algorithm (SMA) is useful as it gives the least error and very good standard deviation. This tells the high reliability and efficiency of SMA over other algorithms. From the convergence curves, we see that there is not much significant advantage of SMA in terms of convergence. The non-parametric assessments, *viz.*, Kruskal-Wallis test and Wilcoxon rank-sum test with added Holm-Bonferroni corrections, also validate the better significance of SMA. We see that the *p*-values are quite low as compared to the threshold value (0.05) taken. Overall, the results obtained from SMA are good and acceptable in comparison to other algorithms. The SMA is useful for the modeling of the Li-ion battery, especially for EV applications where the range of computation time and accuracy is large. Although the first-order ECM captures the battery behavior considerably well, it is not able to furnish any useful information regarding the lifetime of the battery. The performance is somewhat below average when the number of nonlinearity increases. Since the convergence of SMA is quite fast, we can include other critical battery parameters one by one and assess the performance of this algorithm. Also, we can hybridize the SMA with other good metaheuristic algorithms like ALO and GWO to yield better results. The SMA is a newly formed heuristic algorithm that can further be modified/improved for better results and suitability to various applications and superior performance.

References

1. F. Perera, “Pollution from Fossil-Fuel Combustion is the Leading Environmental Threat to Global Pediatric Health and Equity: Solutions Exist.,” *Int. J. Environ. Res. Public Health*, vol. 15, no. 1, December 2017.
2. S. V. Anand, “Global Environmental Issues,” Open Access *Sci. Rep.*, January 2013.
3. J. Larminie and J. Lowry, *Electric Vehicle Technology Explained*, 2nd ed., John Wiley & Sons Ltd, 2012.
4. K. V. Vidyanandan, “Batteries for Electric Vehicles,” November 2019.
5. C. Iclodean, B. Varga, N. Burnete, D. Cimerdean and B. Jurchiş, “Comparison of Different Battery Types for Electric Vehicles,” *IOP Conf. Ser.: Mater. Sci. Eng.*, vol. 252, p. 012058, October 2017.
6. L. Gao, S. Liu and R. A. Dougal, “Dynamic lithium-ion Battery model for system simulation,” Components and Packaging Technologies, *IEEE Trans. on*, vol. 25, pp. 495-505, October 2002.

7. M. Brandl, H. Gall, M. Wenger, V. Lorentz, M. Giegerich, F. Baronti, G. Fantechi, L. Fanucci, R. Roncella, R. Saletti, S. Saponara, A. Thaler, M. Cifrain and W. Prochazka, "Batteries and battery management systems for electric vehicles," in *2012 Design, Automation Test in Europe Conference Exhibition* (DATE), 2012.
8. G. L. Plett, "*LiPB Dynamic Cell Models for Kalman-Filter SOC Estimation*," 2002.
9. G. L. Plett, "Extended Kalman filtering for battery management systems of LiPB-based HEV battery packs Part 2. Modeling and identification," *J. Power Sources*, vol. 134, pp. 262-276, 2004.
10. B. Y. Liaw, G. Nagasubramanian, R. Jungst and D. Doughty, "Modeling of lithium ion cells — A simple equivalent-circuit model approach," *Solid State Ionics*, vol. 175, pp. 835-839, January 2004.
11. K. Hariharan and V. Kumar, "A nonlinear equivalent circuit model for lithium ion cells," *J. Power Sources*, vol. 222, p. 210–217, January 2013.
12. D. Andre, M. Meiler, K. Steiner, H. Walz, T. Soczka-Guth and D. U. Sauer, "Characterization of high-power lithium-ion batteries by electrochemical impedance spectroscopy. II: Modelling," *J. Power Sources*, vol. 196, pp. 5349-5356, 2011.
13. X. Hu, S. Li and H. Peng, "A comparative study of equivalent circuit models for Li-ion batteries," *J. Power Sources*, vol. 198, p. 359–367, January 2012.
14. P. Kumar and P. Bauer, "Parameter extraction of battery models using multiobjective optimization genetic algorithms," in *Proceedings of 14th International Power Electronics and Motion Control Conference EPE-PEMC 2010*, 2010.
15. V. Sangwan, R. Kumar and A. K. Rathore, "Estimation of battery parameters of the equivalent circuit model using Grey Wolf Optimization," in *2016 IEEE 6th International Conference on Power Systems (ICPS)*, 2016.
16. S. Mirjalili, S. M. Mirjalili and A. Lewis, "Grey Wolf Optimizer," *Adv. Eng. Software*, vol. 69, pp. 46-61, 2014.
17. D. H. Wolpert and W. G. Macready, "No free lunch theorems for optimization," *IEEE Trans. Evol. Comput.*, vol. 1, pp. 67-82, 1997.
18. S. Li, H. Chen, M. Wang, A. A. Heidari and S. Mirjalili, "Slime mould algorithm: A new method for stochastic optimization," *Future Gener. Comput. Syst.*, vol. 111, pp. 300-323, 2020.
19. K. Thirugnanam, J. T. P. E. Reena, M. Singh and P. Kumar, "Mathematical Modeling of Li-Ion Battery Using Genetic Algorithm Approach for V2G Applications," *IEEE Trans. Energy Convers.*, vol. 29, pp. 332-343, 2014.
20. M. Chen and G. on-Mora, "Accurate Electrical Battery Model Capable of Predicting Runtime and I–V Performance," IEEE Transactions on Energy Conversion - *IEEE TRANS. ENERGY CONVERS.*, vol. 21, pp. 504-511, June 2006.
21. D. Kessler, "Plasmodial structure and motility," *Cell biology of Physarum and Didymium* / edited by Henry C. Aldrich, John W. Daniel, 1982.

22. T. Nakagaki, H. Yamada and T. Ueda, "Interaction between cell shape and contraction pattern in the Physarum plasmodium.," *Biophys. Chem.*, vol. 84, no. 3, pp. 195-204, May 2000.
23. S. S. MJ Campbell, "*Parametric and Non-parametric tests for comparing two or more groups*," [Online] https://www.healthknowledge.org.uk/public-health-textbook/research-methods/1b-statistical-methods/parametric-non-parametric-tests, 2020.
24. F. Wilcoxon, S. K. Katti and R. A. Wilcox, *Critical values and probability levels for the Wilcoxon rank sum test and the Wilcoxon signed rank test* / [by], [Pearl River, N.Y.]: American Cyanamid, 1963.
25. S. A. McLeod, "What a p-value tells you about statistical significance," *Simply psychology*, pp. 1-4, 2019.
26. Y. O. S. E. F. HOCHBERG, "A sharper Bonferroni procedure for multiple tests of significance," *Biometrika*, vol. 75, pp. 800-802, December 1988.
27. N. Breslow, "A Generalized Kruskal-Wallis Test for Comparing K Samples Subject to Unequal Patterns of Censorship," *Biometrika*, vol. 57, p. 579–594, 1970.
28. "EIG (2000)," [Online] http://www.eigbattery.com, 2000.
29. D. I. MacKenzie, J. D. Nichols, J. A. Royle, K. H. Pollock, L. L. Bailey and J. E. Hines, "Chapter 3 - Fundamental Principals of Statistical Inference," in *Occupancy Estimation and Modeling* (Second Edition), Second Edition ed., D. I. MacKenzie, J. D. Nichols, J. A. Royle, K. H. Pollock, L. L. Bailey and J. E. Hines, Eds., Boston, Academic Press, 2018, pp. 71-111.
30. J. M. Cimbala, "Hypothesis Testing," *Hypothesis Testing*, pp. 1-5, 2014.

12

Harmonic Distortion-Based Performance Analysis and Fault Diagnosis of Inverter Connected with BLDC Motor Using Starting Transients

Surajit Chattopadhyay*, Chiranjit Sain, Purnendu Burui, Sk Rased Ali and Soumya Saha

Dept. of Electrical Engineering, Ghani Khan Choudury, Institute of Engineering and Technology, Malda, West Bengal, India

Abstract

This work deals with "harmonic distortion-based performance analysis and fault diagnosis of inverter connected with brushless DC (BLDC) motor using starting transients." In the field of electrical engineering, advancement has been observed almost in all fields like machinery, inverter technology, grid technology, and control methodology. Simultaneously, changes have been observed in material science, power electronic devices, sensor technology, communications, etc. In this era, any utility system depends on multi-disciplinary units and technologies. Therefore, the performance of a device or unit, or system depends on not only itself but also on various aspects. In this respect, the performance of the inverter also depends on various aspects. In addition, faults may occur outside the system as well as inside the system. It has a become a great challenge to monitor the performance of a system and to detect faults effectively and accurately. Here, an attempt has been taken to find some solutions for performance analysis and fault detection conveniently and accurately. Inverters of various configurations consisting of different power electronics devices connected with solar photovoltaic resources and BLDC motors have been considered as a system of study. The output of the inverter unit has been captured by the data acquisition system (DAS). Current signals are then analyzed by the fast Fourier transform (FFT). The performance analysis and fault diagnosis were carried out. A comparative study was performed that shows different results

**Corresponding author*: surajitchattopadhyay@gmail.com

Tamal Roy, Suman Lata Tripathi and Souvik Ganguli (eds.) Linear and Nonlinear System Modeling, (197–214) © 2024 Scrivener Publishing LLC

for different Inverter configurations. The modeling and simulation performed for various types of inverters like insulated-gate bipolar transistor (IGBT), and metal-oxide-semiconductor field-effect transistor (MOSFET)-based inverter units have shown satisfactory outcomes both in healthy and faulty conditions. Starting transient-based total harmonic distortion (THD) shows significant variations in fault condition from the healthy condition in those inverters. Also, intermediate fault conditions can be discriminated against THD values. In the end, major achievements and scopes for extension of this work have been presented.

Keywords: Brushless DC motor, fault diagnosis, harmonic distortion, inverter switch, performance analysis, semiconductor switch, starting transients, total harmonic distortion

12.1 Introduction

a. Need

In the last few decades, technology has advanced in several directions. In the field of electrical engineering, advancement has been observed almost in all fields like machinery, inverter technology, grid technology, and control methodology. Simultaneously, changes have been observed in material science, power electronic devices, sensor technology, communications, etc. In this era, any utility system depends on multi-disciplinary units and technologies. Therefore, the performance of a device or unit or system depends on not only itself but also on various aspects. In this respect, the performance of the inverter also depends on various aspects. In addition, faults may occur outside the system as well as inside the system. It has become a great challenge to monitor the performance of a system and to detect faults effectively and accurately.

Along with the development of engineering and technology, various mathematical tools have been introduced by different researchers in the last few decades. Some of them have been observed very effective in analyzing images, signals, etc.

Therefore, it has become necessary to find some solutions to monitor performance, detect faults, and cope with other various challenges.

b. Recent Research

Inverter technology and signal assessment have gone through tremendous advancement. At the beginning of this work, a detailed literature survey was carried out. Major observations are presented below.

M. Daniel Pradeep (2015) explained the voltage control method to control the speed of the brushless DC (BLDC) motor [1]. Abhishek Jain (2015) provided a detailed description of an instrumentation technique useful for speed control of a permanent magnet brushless direct current (PM-BLDC) motor [2]. Yasser Ali Almatheel (2017) used fuzzy logic for speed control [3]. Pranoti S.Chaudhari *et al.* (2015) implemented a BLDC motor in a water pumping system [4]. Vinod KR Singh Patel *et al.* (2013) carried out a performance analysis of different proportional–integral–derivative (PID)-controlled BLDC motors [5]. R. Civilian *et al.* (1995) introduced a spindle drive system [6].

Ching-Tsai *et al.* (2015) performed a simulation for torque control using MATLAB [7]. Sanita C S *et al.* (2012) used the hysteresis feature for torque control [8]. P.Nagasekhar Reddy (2013) analyzed the proportional–integral (PI) control system used for the BLDC motor [9]. Salih Baris Ozturk (2011) used both direct and indirect methods to control the speed of the BLDC motor [10]. K. Kamalapathi *et al.* (2015) used torque control for the BLDC motor [11].

T. Chun *et al.* (2014) introduced a BLDC motor drive for pumping purposes by comparison of hysteresis based on sensorless control [13]. A. Usman *et al.* (2019) introduced a demagnetization fault in BLDC motor by numerical based as well as using hybrid electrical circuit [14]. W. Li *et al.* (2016) introduced a sensorless position control high-speed BLDC motor that acts without phase shift [15].

M. Baszynski *et al.* (2018) carried out closed-loop control for a BLDC motor with unipolar modulation [16]. G. Liu *et al.* (2017) introduced a sensorless BLDC motor with a low start-up current for small inductance [17]. S. Chen *et al.* (2018) carried out a position correction-based sensorless BLDC motor that has low stator inductance [18]. R. Horvat *et al.* (2014) introduced an field programmable gate arrays (FPGA)-based BLDC motor using event-driven current control [19]. X. Zhou *et al.* (2017) introduced a low-induction BLDC motor that compensates by itself using rapid self-compensation for commutation phase error [20]. S. Neethu *et al.* (2014) introduced a hysteresis control BLDC motor that has an open phase fault controller [21].

S. K. Chari *et al.* (2018) introduced a simulation-based hysteresis current control inverter design that has been used for BLDC motor [22]. H. Suryoatmojo *et al.* (2018) carried out the current control mechanism using hysteresis to control the speed of a BLDC motor [23]. Y. K. Lee *et al.* (2019) made an attempt to minimize the torque ripple and switching power loss of a BLDC motor [24]. Ebin Joseph T *et al.* (2015) performed direct torque control to achieve better speed control of the drive [25]. J. E. Muralidhar *et al.* (2014) introduced closed-loop control to minimize torque ripple [26]. M. S. Trivedi *et al.* (2017) introduced predictive control using hysteresis current [27].

C. Chhlonh *et al.* (2019) used the fuzzy controller to control the speed of the rear wheel using BLDC [28]. A. Azarudeen *et al.* (2017) analyzed the performance of the pulse width modulation (PWM) controller used for the BLDC motor [29]. M. Ebadpour *et al.* (2017) introduced a model for dual BLDC motor drives connected in parallel [30]. M. Nicola *et al.* (2019) utilized a back electromotive force (EMF) observer to achieve sensorless control of the motor [31]. A. Usman *et al.* (2019) made a detailed analysis of faults that occurred due to the demagnetization effect [32].

N. Seth *et al.* (2017) made a detailed review on inverter topologies that are useful for industries [33]. Tomoya Sugimoto *et al.* (2019) applied T type boost inverter that gives better performance with the help of switching capacitors [34]. Tomoya Sugimoto *et al.* (2018) reduced switching loss using multilevel topology with multiple numbers of inverters [35] Pablo Montero Robina *et al.* (2018) utilized a converter with a five-level diode combination to interface with the solar grid [36].

S. Bernet *et al.* (1999) compared the performance of different types of inverters [37]. H. Miyazaki *et al.* (2000) used IGBT-based clamped inverter for industry applications [38]. H. A. Mantooth *et al.* (1997) made a simulation of a PWM inverter considering electro-thermal features [39]. A. Ramamurthy *et al.* (1999) modeled voltage change during switching off the condition for IGBT applications [40].

B. Chen *et al.* (2015) modeled MOSFET-based microinverter with high efficiency [41]. O. Sivkov *et al.* (2018) compared silicon-based inverters [42]. H. Kernstock *et al.* (2012) used soft switching methods to improve efficiency [43]. S. Mangkalajarn *et al.* (2019) used class-E inverter to achieve optimum performance [44].

Literature survey shows that a lot of works have been carried out to model, design, and to analyze the performance of different converters used in drives' applications. However, very few works have been noticed to utilize harmonics patterns for performance analysis and fault diagnosis. This has motivated the authors to find some solutions for performance analysis and fault detection conveniently and accurately. The inverter of various configurations consisting of different power electronics devices connected with solar photovoltaic (PV) resource and BLDC motor have been considered as a system of study.

c. Challenges

The challenges and needs mentioned in the previous sections have motivated the work team to find some solutions for performance analysis and fault detection conveniently and accurately. The inverter of various configurations consisting of different power electronic devices connected with solar PV resource and BLDC motor have been considered as a system of study. Finally, the work aims to do the following:

a. To model the inverter of various configurations consisting of different power electronics devices connected with solar PV resource and BLDC motor
b. To monitor the performance of inverter in terms of harmonic distortion
c. To detect fault that often occurs at power electronic switches inside an inverter

12.2 Modeling

With the advancement of various types of inverter technology and the advancement of electrical machinery, the use of inverter-connected motor applications has increased to a great extent. Therefore, it has become a necessary and urgent job to monitor the performance of the inverter-driven motor unit.

In this work, an attempt has been taken to monitor the performance of a solar PV-driven inverter that feeds power to a BLDC motor and to detect short circuit faults inside the inverter unit.

Modeling was done by using MATLAB (R2018a) with different types of inverter configurations. This analysis was done in two conditions, as follows:

I. Normal condition (assuming on-state resistance $R = 0.001\ \Omega$)
II. Fault condition (by multiplying the on-state by 10, from $R = 0.01$ to $R = 10^6\ \Omega$)

PowerGUI tool was used to carry out fast Fourier transform (FFT) analysis of the phase currents.

a. Inverter- BLDC Motor Model With Hysteresis Current Control

The block diagram of the BLDC motor with hysteresis current control is shown in Figures 12.1, 12.2, and 12.3. It consists of four major units, as follows:

1. Energy unit
2. Converter unit
3. Motor unit
4. Data capture unit

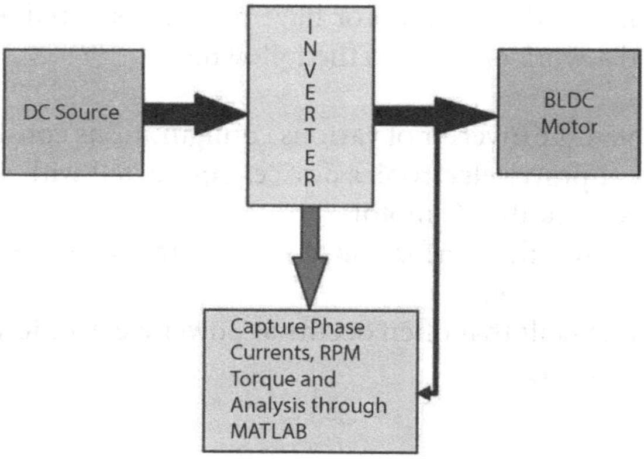

Figure 12.1 Block diagram of BLDC motor with hysteresis current control.

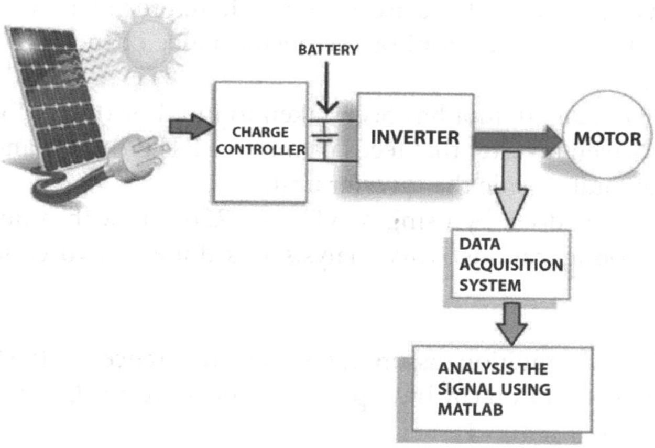

Figure 12.2 Block diagram of solar PV-connected BLDC motor model.

b. Energy Unit

The energy unit consists of a battery unit. The unit is charged by solar PV resources through the charge controller.

c. Converter Unit

The converter unit consists of inverters to provide a three-phase supply to the motor unit. Mainly, there are three types of inverter configurations that were considered for modeling and simulation.

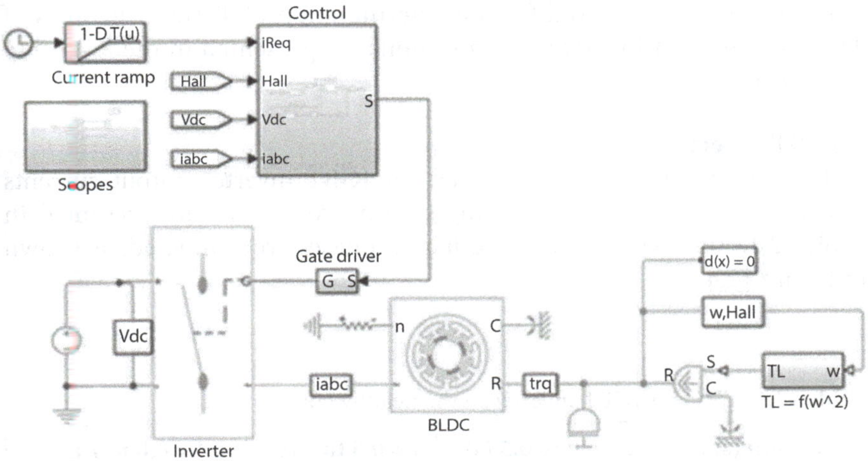

Figure 12.3 Modeling of BLDC motor with hysteresis current control (MATLAB 18).

d. Motor and Data Capture Unit

In this, the BLDC motor was considered as load.

The output current of the inverter unit was captured by the data acquisition system (DAS). Current signals were then analyzed by FFT. The performance analysis and fault diagnosis were carried out.

The sub-part consists of three-phase current monitoring scopes, revolutions per minute (RPM) and torque monitoring units. A power GUI tool was added to analyze the FFT of current signals.

e. Layout Diagram of Proposed Work

The block diagram of the solar PV-connected BLDC motor model is shown in Figure 12.2. It consists of solar PV resources connected to the battery unit charge controller. The charge controller is connected to the battery. The battery is connected to the inverter and supplies power to the BLDC motor through the inverter. The inverter's output current is captured through the data acquisition system and analyzed to monitor the performance of the inverter-motor configuration and to detect faults in the inverter unit.

12.3 THD Comparison of Phase Currents of Different Inverters

Harmonic distortion for different captured current signals was determined for different inverter configurations. Three different time durations were considered for the determination of THD. Outcomes of THD for different inverter configurations are presented in the following subsections.

a. IGBT Inverter

Total harmonic distortion (THD) for IGBT inverter output currents drawn by the BLDC motor during starting conditions are presented in Table 12.1. The THD values for different phases are compared, as shown in Figure 12.4.

Table 12.1 THD for IGBT inverter in the healthy condition.

Duration (sec)	t = 0.0 to 0.5 (s)	t = 0.5 to 1 (s)	t = 0.0 to 1 (s)
THD of ia	518.7	6086.57	1756.69
THD of ib	538.24	6016.06	1841.55
THD of ic	531.7	6413.79	1803.47

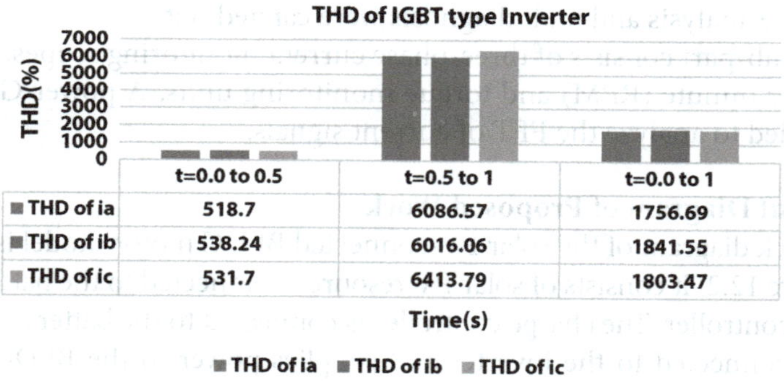

Figure 12.4 THD for IGBT inverter in the healthy condition.

Table 12.2 THD for MOSFET inverter in the healthy condition.

Duration (s)	t = 0.0 to 0.5 (s)	t = 0.5 to 1 (s)	t = 0.0 to 1 (s)
THD of ia	553.09	11,977.67	1690.01
THD of ib	549.47	14,690.87	1530.75
THD of ic	525.41	7118.72	1682.7

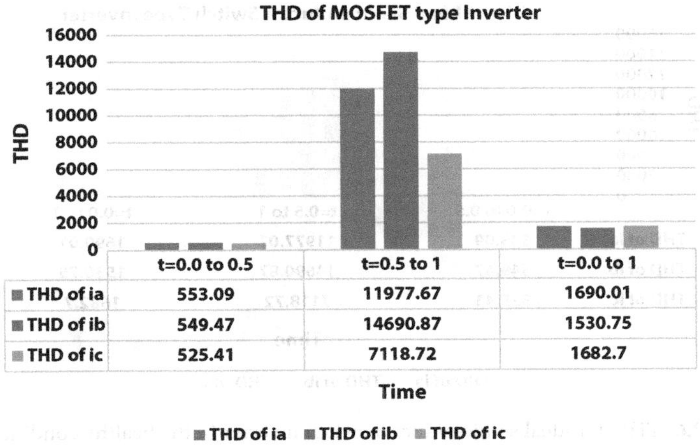

Figure 12.5 THD for MOSFET inverter in the healthy condition.

b. MOSFET Inverter

The THD for MOSFET inverter output currents drawn by the BLDC motor during starting conditions is presented in Table 12.2. The THD values for different phases are compared, as shown in Figure 12.5.

c. Ideal Semiconductor Switch Inverter

The THD for the 'ideal semiconductor switch inverter' output currents drawn by the BLDC motor during starting conditions are presented in Table 12.3. The THD values for different phases are compared, as shown in Figure 12.6.

d. Comparative Study and Specific Outcome

Harmonic distortions (THDs) of different configurations of the inverter units were compared for different time durations. The results of the comparison are presented in Figures 12.7 to 12.9. The THD comparison of different types of inverters for t = 0.0 s to 0.5 s are shown in Figure 12.7.

Table 12.3 THD for ideal semiconductor switch inverter in the healthy condition.

Duration	t = 0.0 to 0.5 (s)	t = 0.5 to 1 (s)	t = 0.0 to 1 (s)
THD of ia	533.09	11977.01	1690.01
THD of ib	549.47	14,690.87	1530.75
THD of ic	525.41	7118.72	1682.7

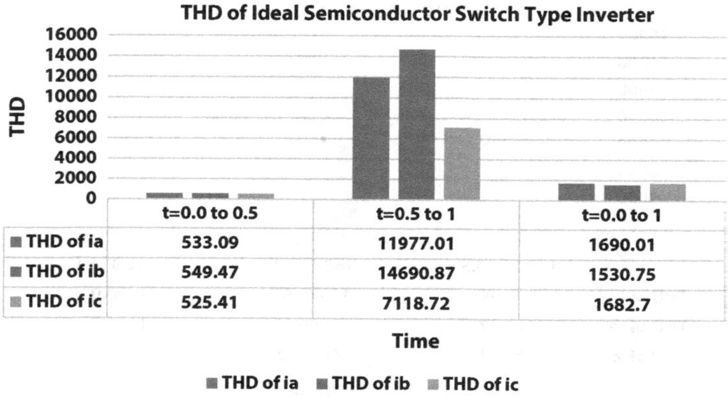

	t=0.0 to 0.5	t=0.5 to 1	t=0.0 to 1
THD of ia	533.09	11977.01	1690.01
THD of ib	549.47	14690.87	1530.75
THD of ic	525.41	7118.72	1682.7

Figure 12.6 THD for ideal semiconductor switch inverter in the healthy condition.

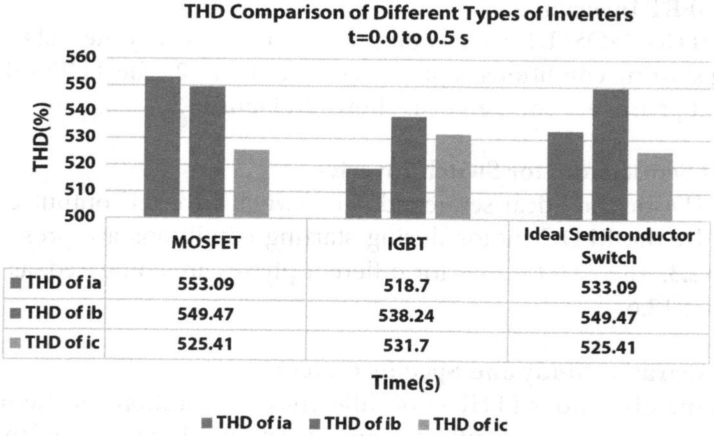

	MOSFET	IGBT	Ideal Semiconductor Switch
THD of ia	553.09	518.7	533.09
THD of ib	549.47	538.24	549.47
THD of ic	525.41	531.7	525.41

Figure 12.7 THD comparison of different types of inverters for t = 0.0 s to 0.5 s.

The THD comparison of different types of inverters for t = 0.5 s to 1 s are shown in Figure 12.8. The THD comparison of different types of inverters for t = 0.0 s to 1 s are shown in Figure 12.9. These bar charts show different harmonic distortions for different inverter configurations. Therefore, harmonic distortion-based performance analysis should have prior knowledge of inverter configuration.

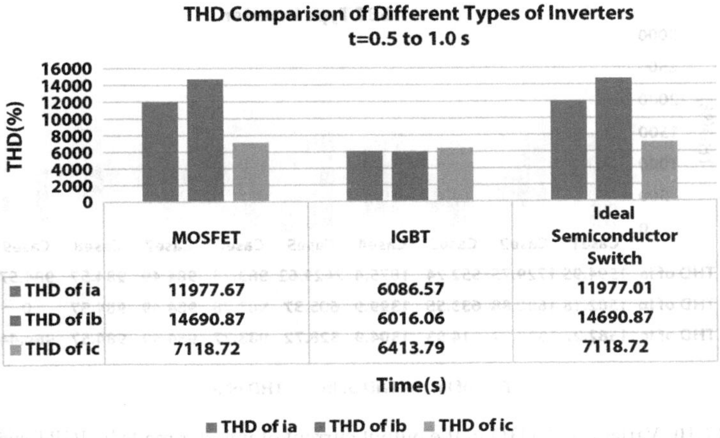

	MOSFET	IGBT	Ideal Semiconductor Switch
THD of ia	11977.67	6086.57	11977.01
THD of ib	14690.87	6016.06	14690.87
THD of ic	7118.72	6413.79	7118.72

Figure 12.8 THD comparison of different types of inverters for t = 0.5 s to 1 s.

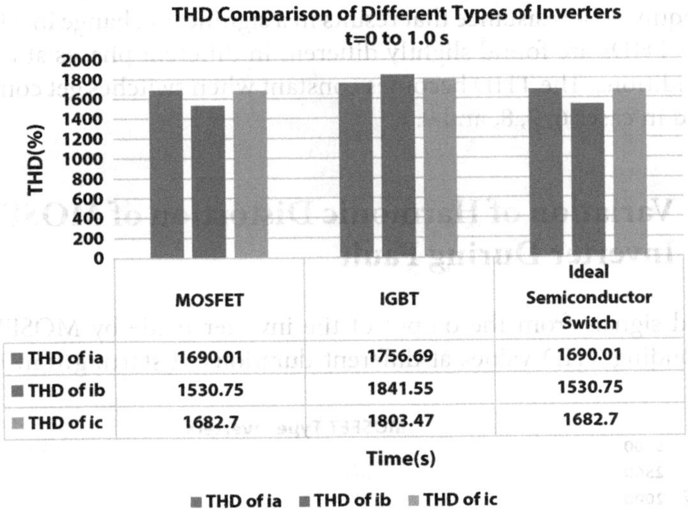

	MOSFET	IGBT	Ideal Semiconductor Switch
THD of ia	1690.01	1756.69	1690.01
THD of ib	1530.75	1841.55	1530.75
THD of ic	1682.7	1803.47	1682.7

Figure 12.9 THD comparison of different types of inverters for t = 0.0 s to 1 s.

12.4 Variation of Harmonic Distortion of IGBT Inverter During Fault

Captured signals from the output of the inverter made by IGBT and corresponding THD values at different durations of starting transient are presented here. From these outcomes, the THD for different intermediate and

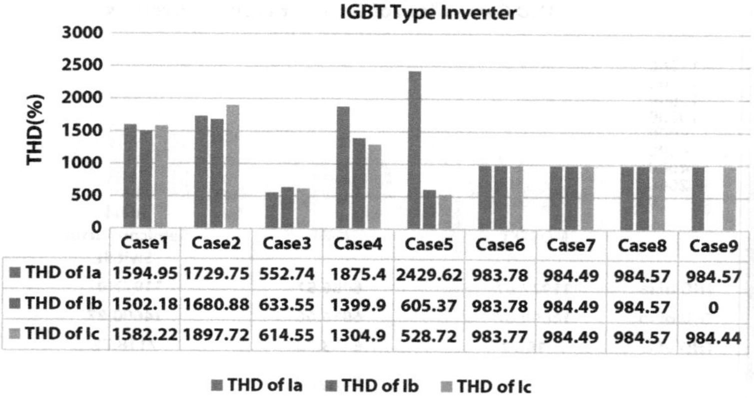

	Case1	Case2	Case3	Case4	Case5	Case6	Case7	Case8	Case9
THD of Ia	1594.95	1729.75	552.74	1875.4	2429.62	983.78	984.49	984.57	984.57
THD of Ib	1502.18	1680.88	633.55	1399.9	605.37	983.78	984.49	984.57	0
THD of Ic	1582.22	1897.72	614.55	1304.9	528.72	983.77	984.49	984.57	984.44

Figure 12.10 Variation of THD in the output current of inverter made by IGBT switches.

extreme faults are noted and presented in Figure 12.10. The figure clearly shows that the fault in the power electronic switch is reflected in the variation of equivalent resistance that results in a significant change in THD values. The THDs are found slightly different in different phases at incipient fault conditions. The THD becomes constant when switches get completely damaged in cases 6, 7, 8, and 9.

12.5 Variation of Harmonic Distortion of MOSFET Inverter During Fault

Captured signals from the output of the inverter made by MOSFET and corresponding THD values at different durations of starting transient are

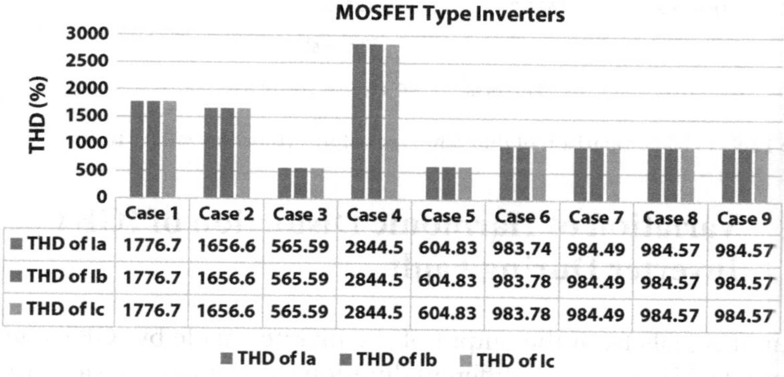

	Case 1	Case 2	Case 3	Case 4	Case 5	Case 6	Case 7	Case 8	Case 9
THD of Ia	1776.7	1656.6	565.59	2844.5	604.83	983.74	984.49	984.57	984.57
THD of Ib	1776.7	1656.6	565.59	2844.5	604.83	983.78	984.49	984.57	984.57
THD of Ic	1776.7	1656.6	565.59	2844.5	604.83	983.78	984.49	984.57	984.57

Figure 12.11 Variation of THD in the output current of inverter made by MOSFET switches.

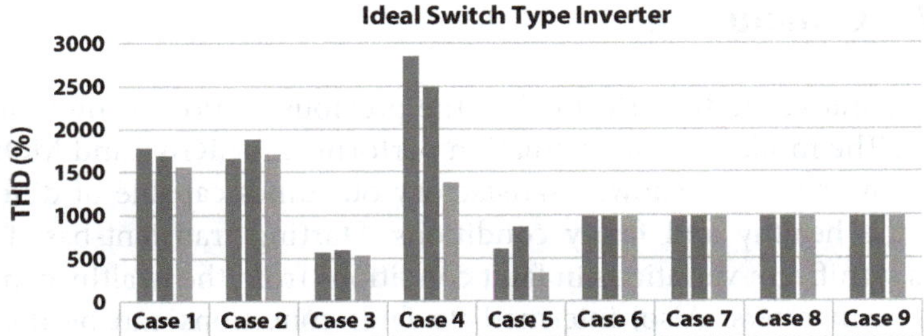

	Case 1	Case 2	Case 3	Case 4	Case 5	Case 6	Case 7	Case 8	Case 9
THD of Ia	1776.72	1656.6	565.59	2844.46	604.83	983.78	984.49	984.57	984.57
THD of Ib	1687.57	1879.42	592.65	2493.63	861.63	983.78	984.49	984.57	984.46
THD of Ic	1554.36	1703.49	530.91	1366.54	326.78	983.77	984.49	984.57	984.46

THD of Ia THD of Ib THD of Ic

Figure 12.12 Variation of THD in the output current of Inverter made by ISS switches.

presented here. From these outcomes, the THD for different intermediate and extreme faults are noted and presented in Figure 12.11. The figure clearly shows that the fault in the power electronic switch is reflected in the variation of equivalent resistance that results in a significant change in THD values. Unlike inverters with IGBT, here, THD values are the same in all phases. During completely damaged condition (equivalent to opening or breaking of the switching path), THD becomes constant in the last four cases.

12.6 Variation of Harmonic Distortion of Ideal Switch Inverter During Fault

Captured signals from the output of the inverter made by ideal semiconductor switch (ISS) and corresponding THD values at different durations of starting transient are presented here. From these outcomes, the THD for different intermediate and extreme faults are noted and presented in Figure 12.12. The figure, clearly shows that the fault in the power electronic switch is reflected in the variation of equivalent resistance that results in a significant change in THD values. The THD variation is found almost the same as IGBT-based inverter. It shows the different THD values during intermediate fault conditions and constant values in extreme conditions in the last four cases.

12.7 Conclusion

A comparative study performed in the previous sections shows different results. The modeling and simulation performed for IGBT and MOSFET-based inverter units showed satisfactory outcomes capable of differentiating the healthy and faulty conditions. Starting transient-based THD shows significant variations in fault conditions from the healthy condition in those inverters. Also, intermediate fault conditions can be discriminated against using THD values. However, ideal switch-based inverter units have shown satisfactory outcomes in healthy conditions and failed to run the motor to provide necessary torque and speed to the mechanical load during fault conditions. Hence, in this work, THD-based fault analysis of the inverter with ideal switch could not be completed using starting transient. Major achievements of this work are as follows: (a) modeling of battery-inverter-BLDC motor system, (b) effective utilization of various inverter configurations with different power electronic switches, (c) capture of starting transient, (d) determination of harmonic distortion and THD values of starting transients at healthy condition, (e) determination of harmonic distortion and THD values of starting transients at faulty condition, (f) comparison of THD values, and (g) discrimination of healthy and faulty conditions. The work may further be extended to carry out the following: (a) the work may be extended for performance analysis and fault diagnosis of motor units, (b) motor other than BLDC can be used in further study, (c) different energy resources can be used in further study, (d) grid-connected inverter unit can be tested, (e) steady-state and switching off transients can be analyzed, (f) other mathematical tools can be used to analyze current signatures, and (g) the work may be extended for analysis and diagnosis from remote end using Internet of Things and the Cloud.

References

1. M. Daniel Pradeep, "A Novel Method of Speed and Voltage Control of BLDC Motor" *IJCSMC*, Vol. 4, Issue. 3, March 2015.
2. Abhishek Jain, "Controlling of Permanent Magnet Brushless Dc Motor Using Instrumentation Technique" *Int. J. Adv. Eng. Res. Dev.* Volume 2, Issue 1, Jan-2015.
3. Yasser Ali Almatheel, "Speed Control of Dc Motor Using Fuzzy Logic Controller" *2017 International Conference on Communication, Control, Computing And*

4. Pranoti S. Chaudhari, 2, Girish K.Mahajan, "Implementation of BLDC Motor Based Water Pump for Automotive Vehicle" *Int J Eng Sci*, Volume 4, Issue 6, PP.34-4, June – 2015.
5. Vinod KR Singh Patel, A.K.Pandey, "Modeling and Performance Analysis of PID Controlled BLDC Motor and Different Schemes of PWM Controlled BLDC Motor "*Int. J. Sci. Res. Publ.*, Volume 3, Issue 4, April 2013.
6. R. Civilian, and D. Stupak, "Disk drive employing multi-mode spindle drive system," US patent 5471353, Oct 3, 1995.
7. Ching-Tsai, P. and C. Ting-Yu, Kamalapathi.K1, Vinod Kumar.P2, Balaji. C3, "TORQUE CONTROL OF BLDC MOTOR DRIVE USING MATLAB/ SIMULINK" *International Journal of Emerging Trends in Engineering Research (IJETER)*, Vol. 3 No.6, Pages: 448 - 452 (2015) Special Issue of NCTET 2K15 - Held on June 13, 2015, in SV College of Engineering, Tirupati. http://warse.org/IJETER/static/pdf/Issue/NCTET2015sp80.pdf
8. Sunita C S1, J T Kuncheria, Ahmad Faiz Noor Azam, "TORQUE HYSTERESIS CONTROLLER FOR BRUSHLESS DC MOTOR DRIVES" *Power and Energy Conversion Symposium (PECS 2012)* Melaka, Malaysia 17 Dec 2012.
9. P. Nagasekhar Reddy, "Modeling and Analysis of PI Controller Based Speed Control of Brushless DC Motor Drive". *Int. J. Eng. Sci. Technol.*, Reddy, 2(9): September 2013.
10. Salih Baris Ozturk, "Direct Torque and Indirect Flux Control of Brushless DC Motor", *IEEE/ASME Trans. Mechatron.*, VOL.16, NO.2, APRIL 2011.
11. Kamalapathi.K1, Vinod Kumar.P2, Balaji.C3, "torque control of BLDC motor drive using MATLAB/SIMULINK" *International Journal of Emerging Trends in Engineering Research (IJETER)*, Vol. 3 No.6, Pages: 448-452 (2015) Special Issue of NCTET 2K15 - Held on June 13, 2015, in SV College of Engineering, Tirupati.
12. Yong Liu, "Direct Torque Control of Brushless DC Drives with Reduced Torque Ripple" *IEEE Trans. Ind. Appl.*, vol. 41, no. 2, March/April 2005.
13. T. Chun, Q. Tran, H. Lee, and H. Kim, "Sensorless Control of BLDC Motor Drive for an Automotive Fuel Pump Using a Hysteresis Comparator," in *IEEE Trans. Power Electron.*, vol. 29, no. 3, pp. 1382-1391, March 2014, DOI: 10.1109/TPEL.2013.2261554.
14. A. Usman and B. S. Rajpurohit, "Comprehensive Analysis of Demagnetization Faults in BLDC Motors Using Novel Hybrid Electrical Equivalent Circuit and Numerical Based Approach," in *IEEE Access*, vol. 7, pp. 147542-147552, 2019, DOI: 10.1109/ACCESS.2019.2946694.
15. W. Li, J. Fang, H. Li and J. Tang, "Position Sensorless Control Without Phase Shifter for High-Speed BLDC Motors With Low Inductance and Nonideal Back EMF," in *IEEE Trans. Power Electron.*, vol. 31, no. 2, pp. 1354-1366, Feb. 2016, DOI: 10.1109/TPEL.2015.2413593.
16. M. Baszynski and S. Pirog, "Unipolar Modulation for a BLDC Motor with Simultaneously Switching of Two Transistors with Closed Loop Control for

Four-Quadrant Operation," in *IEEE Trans. Ind. Inf.*, vol. 14, no. 1, pp. 146-155, Jan. 2018, DOI: 10.1109/TII.2017.2723962.

17. G. Liu, S. Chen, S. Zheng and X. Song, "Sensorless Low-Current Start-Up Strategy of 100-kW BLDC Motor with Small Inductance," in *IEEE Trans. Ind. Inf.*, vol. 13, no. 3, pp. 1131-1140, June 2017, DOI: 10.1109/TII.2016.2607158.
18. S. Chen, W. Sun, K. Wang, G. Liu, and L. Zhu, "Sensorless High-Precision Position Correction Strategy for a 100 kW@20 000 r/min BLDC Motor with Low Stator Inductance," in *IEEE Trans. Ind. Inf.*, vol. 14, no. 10, pp. 4288-4299, Oct. 2018, DOI: 10.1109/TII.2018.2793947.
19. R. Horvat, K. Jezernik and M. Čurkovič, "An Event-Driven Approach to the Current Control of a BLDC Motor Using FPGA," in *IEEE Trans. Ind. Electron.*, vol. 61, no. 7, pp. 3719-3726, July 2014, DOI: 10.1109/TIE.2013.2276776.
20. X. Zhou, X. Chen, M. Lu and F. Zeng, "Rapid Self-Compensation Method of Commutation Phase Error for Low- Inductance BLDC Motor," in *IEEE Trans. Ind. Inf.*, vol. 13, no. 4, pp. 1833-1842, Aug. 2017, DOI: 10.1109/TII.2017.2653812.
21. S. Neethu and V. Sreelekha, "Hysteresis controller based open phase fault tolerant control of BLDC motor drives," *2014 International Conference on Power Signals Control and Computations (EPSCICON)*, 2014, pp. 1-6, DOI: 10.1109/EPSCICON.2014.6887514.
22. S. K. Chari, R. Dhiman and R. Saxena, "Novel and Robust Hysteresis Current Control Strategies for a BLDC Motor: A Simulation Study and Inverter Design," *2018 2nd IEEE International Conference on Power Electronics, Intelligent Control and Energy Systems (ICPEICES)*, 2018, pp. 841-845, DOI: 10.1109/ICPEICES.2018.8897424.
23. H. Suryoatmojo, F. G. Cladella, V. Lystianingrum, D. C. Riawan, R. Mardiyanto, and D. Ariana, "Performance of BLDC Motor Speed Control Based on Hysteresis Current Control Mechanism," *2018 International Seminar on Intelligent Technology and Its Applications (ISITIA)*, 2018, pp. 147-152, DOI: 10.1109/ISITIA.2018.8710910.
24. Y. K. Lee, "Torque Ripple and Switching Power Loss Minimization with Constant Band Hysteresis Current Controller for BLDC Motor," *2019 IEEE PES Asia-Pacific Power and Energy Engineering Conference (APPEEC)*, 2019, pp. 1-4, DOI: 10.1109/APPEEC45492.2019.8994737.
25. Ebin Joseph T and Sreethumol M V, "Speed control of BLDC motor drive under Direct Torque Control scheme with modified integrator," *2015 IEEE International Conference on Electrical, Computer and Communication Technologies (ICECCT)*, 2015, pp. 1-5, DOI: 10.1109/ICECCT.2015.7225931.
26. J. E. Muralidhar and P. V. Aranasi, "Torque ripple minimization & closed-loop speed control of BLDC motor with hysteresis current controller," *2014 2nd International Conference on Devices, Circuits and Systems (ICDCS)*, 2014, pp. 1-7, DOI: 10.1109/ICDCSyst.2014.6926122.
27. M. S. Trivedi, R. K. Keshri and V. Ravikiran, "Predictive hysteresis current control of PM BLDC motor in stationary plane," *2017 IEEE Transportation*

Electrification Conference (ITEC-India), 2017, pp. 1-5, DOI: 10.1109/ITEC-India.2017.8356941.

28. C. Chhlonh, D. C. Riawan, and H. Suryoatmojo, "Independent Speed Steering Control of Rear In-wheel BLDC Motor in EV Based on Fuzzy Logic Controller in GUI," *2019 5th International Conference on Science and Technology (ICST)*, 2019, pp. 1-6, DOI: 10.1109/ICST47872.2019.9166418.
29. A. Azarudeen and D. Mary, "Performance analysis of conventional and digital PWM control scheme for speed control of BLDC motor drives," *2017 International Conference on Advances in Electrical Technology for Green Energy (ICAETGT)*, 2017, pp. 69-75, DOI: 10.1109/ICAETGT.2017.8341460.
30. M. Ebadpour, M. B. B. Sharifian, and E. Babaei, "Modeling and control of dual parallel BLDC motor drive system with single inverter," IECON 2017 - *43rd Annual Conference of the IEEE Industrial Electronics Society*, 2017, pp. 3740-3743, DOI: 10.1109/IECON.2017.8216636.
31. M. Nicola, C. Nicola, and D. Sacerdoţianu, "Sensorless Control of Multi-Motors BLDC using Back-EMF Observer," *2019 8th International Conference on Modern Power Systems (MPS)*, 2019, pp. 1-6DOIoi: 10.1109/MPS.2019.8759761.
32. A. Usman and B. S. Rajpurohit, "Comprehensive Analysis of Demagnetization Faults in BLDC Motors Using Novel Hybrid Electrical Equivalent Circuit and Numerical Based Approach," in *IEEE Access*, vol. 7, pp. 147542-147552, 2019, DOI: 10.1109/ACCESS.2019.2946694.
33. N. Seth, V. Goel, and R. D. Kulkami, "Three-phase innovative multilevel inverter topologies for research and industrial applications: A review," *2017, International Conference on Nascent Technologies in Engineering (ICNTE)*, 2017, pp. 1-6, DOI: 10.1109/ICNTE.2017.7947934.
34. Tomoya Sugimoto, Takahiro Nozaki, Toshiyuki Murakami, "Extended T-Type Boost Inverter Using Switched Capacitors", *IECON 2019 - 45th Annual Conference of the IEEE Industrial Electronics Society*, vol.1, pp. 1756-1761, 2019.
35. Tomoya Sugimoto, Takahiro Nozaki, Toshiyuki Murakami, "Multilevel Inverter Topology for Switching Loss Reduction", *IECON 2018 - 44th Annual Conference of the IEEE Industrial Electronics Society*, pp. 1345-1350, 2018.
36. Pablo Montero Robina, Francisco Gordillo, "A Novel Controller for Grid-Interfacing Solar Arrays Through Five-Level Diode-Clamped Converters", *2018 IEEE 18th International Power Electronics and Motion Control Conference (PEMC)*, pp. 955-961, 2018.
37. S. Bernet, R. Teichmann, A. Zuckerberger and P. K. Steimer, "Comparison of high-power IGBT's, and hard-driven GTOs for high-power inverters," in *IEEE Trans. Ind. Appl.*, vol. 35, no. 2, pp. 487-495, March-April 1999, DOI: 10.1109/28.753645.
38. H. Miyazaki, H. Fukumoto, S. Sugiyama, M. Tachikawa, and N. Azusawa, "Neutral-point-clamped inverter with parallel driving of IGBTs for industrial applications," in *IEEE Trans. Ind. Appl.*, vol. 36, no. 1, pp. 146-151, Jan.-Feb. 2000, DOI: 10.1109/28.821809.

39. H. A. Mantooth and A. R. Hefner, "Electrothermal simulation of an IGBT PWM inverter," in *IEEE Trans. Power Electron.*, vol. 12, no. 3, pp. 474-484, May 1997, DOI: 10.1109/63.575675.
40. A. Ramamurthy, S. Sawant, and B. J. Baliga, "Modeling the [dV/dt] of the IGBT during inductive turn-off," in *IEEE Trans. Power Electron.*, vol. 14, no. 4, pp. 601-606, July 1999, DOI: 10.1109/63.774195.
41. B. Chen et al., "A High-Efficiency MOSFET Transformerless Inverter for Nonisolated Microinverter Applications," in *IEEE Trans. Power Electron.*, vol. 30, no. 7, pp. 3610-3622, July 2015, DOI: 10.1109/TPEL.2014.2339320.
42. O. Sivkov, M. Novak and J. Novak, "Comparison between Si IGBT and SiC MOSFET Inverters for AC Motor Drive," *2018 18th International Conference on Mechatronics - Mechatronika (ME)*, 2018, pp. 1-5.
43. H. Kernstock and B. Plassnegger, "High efficiency soft switched 3-level MOSFET Inverter for an Electric Vehicle PMSM Drive," *2012 15th International Power Electronics and Motion Control Conference (EPE/PEMC)*, 2012, pp. DS1b.9-1-DS1b.9-5, DOI: 10.1109/EPEPEMC.2012.6397210.
44. S. Mangkalajarn, C. Ekkaravarodome, S. Sukanna, A. Bilsalam, K. Jirasereeamongkul, and K. Higuchi, "Comparative Study of Si IGBT and SiC MOSFET in Optimal Operation Class-E Inverter for Domestic Induction Cooker," *2019 Research, Invention, and Innovation Congress (RI2C)*, 2019, pp. 1-4, DOI: 10.1109/RI2C48728.2019.8999967.

About the Editors

Tamal Roy, PhD, received his PhD from Jadavpur University in 2016. In 2008, he joined the Department of Electrical Engineering at Hooghly Engineering and Technology College as a Lecturer with 15 years of academic experience. Since 2011, he has been working as an assistant professor in the Electrical Engineering Department of the MCKV Institute of Engineering and presently is Head of the Department. His current research interests include adaptive control, uncertainty modelling, and robust control of nonlinear systems.

Suman Lata Tripathi, PhD is a professor at the Lovely Professional University with more than 20 years of experience in academics. She is also a remote post-doctoral researcher at Nottingham Trent University, London, UK. She has published more than 74 research papers in refereed science journals and conferences, as well as 13 Indian patents and two copyrights. Additionally, she has edited and authored more than 17 books in different areas of electronics and electrical engineering.

Souvik Ganguli, PhD is associated with the Thapar Institute of Engineering and Technology, Patiala as an assistant professor since June 2009 with fourteen years of experience in academics. Before joining academics he served the industry for more than two years. He has published eight Science Citation Index journal papers and nearly 50 Scopus indexed papers, book chapters, and conferences. Recently, he has been granted an Australian Innovation patent for his contribution to the industrial cyber-physical system and eight of his patents are already published and awaiting grants.

Index

Also of Interest

By the same editors

INTEGRATED DEVICES FOR ARTIFICIAL INTELLIGENCE AND VLSI: VLSI Design, Simulation and Applications, Edited by Balwinder Raj, Suman Lata Tripathi, Tarun Chaudhary, K. Srinivasa Rao, and Mandeep Singh, ISBN: 9781394204359. With its in-depth exploration of the close connection between microelectronics, AI, and VLSI technology, this book offers valuable insights into the cutting-edge techniques and tools used in VLSI design automation, making it an essential resource for anyone seeking to stay ahead in the rapidly evolving field of VLSI design.

DECENTRALIZED SYSTEMS AND DISTRIBUTED COMPUTING, Edited by Sandhya Avasthi, Suman Lata Tripathi, Namrata Dhanda, and Satya Bhushan Verma, ISBN: 978139420436. This book provides a comprehensive exploration of next-generation internet, distributed systems, and distributed computing, offering valuable insights into their impact on society and the future of technology.

ELECTRIC VEHICLE DESIGN: Design, Simulation and Applications, Edited by Krishan Arora, Suman Lata Tripathi, and Himashu Sharma, ISBN: 9781394204373. This book will serve as a definitive guide to conceptual and practical knowledge about the design of hybrid electrical vehicles (HEV), battery electrical vehicles (BEV), fuel cell electrical vehicles (FCEV), plug-in hybrid electrical vehicles (PHEV), and efficient EV charging techniques with advanced tools and methodologies for students, engineers, and academics alike.

INDUSTRIAL CONTROL SYSTEMS, Edited by Vipin Chandra Pal, Suman Lata Tripathi, and Souvik Ganguli, ISBN: 9781119829256. This volume serves as a comprehensive guide in the journey of industrial control systems with a multidisciplinary approach to the key engineering problems in the 21st century.

MODERN AUTOMOTIVE ELECTRICAL SYSTEMS: Theory and Applications, Edited by Pedram Asef, Sanjeevikumar Padmanaban, and Andrew Lapthorn, ISBN: 1119801047. Presenting the concepts and advances of modern automotive electrical systems, this volume, written and edited by a global team of experts, also goes into the practical applications for the engineer, student, and other industry professionals.

NANODEVICES FOR INTEGRATED CIRCUIT DESIGN, Edited by Suman Lata Tripathi, Abhishek Kumar, K. Srinivasa Rao, and Prasantha R. Mudimela, ISBN: 9781394185788. Written and edited by a team of experts in the field, this important new volume broadly covers the design of nano-devices and their integrated applications in digital and analog integrated circuits (IC) design.

EXPLAINABLE MACHINE LEARNING MODELS AND ARCHITECTURES: Real-Time System Implementation, Edited by Suman Lata Tripathi and Mufti Mahmud, ISBN: 9781394185849. This cutting-edge new volume covers the hardware architecture implementation, the software implementation approach, and the efficient hardware of machine learning applications.

MACHINE LEARNING TECHNIQUES FOR VLSI CHIP DESIGN, Edited by Abhishek Kumar, Suman Lata Tripathi, and K. Srinivasa Rao, ISBN: 9781119910398. This cutting-edge new volume covers the hardware architecture implementation, the software implementation approach, and the efficient hardware of machine learning applications with FPGA or CMOS circuits, and many other aspects and applications of machine learning techniques for VLSI chip design.

INTELLIGENT GREEN TECHNOLOGIES FOR SMART CITIES, Edited by Suman Lata Tripathi, Souvik Ganguli, Abhishek Kumar, and Tengiz Magradze, ISBN: 9781119816065. Presenting the concepts and fundamentals of smart cities and developing "green" technologies, this volume, written and edited by a global team of experts, also goes into the practical applications that can be utilized across multiple disciplines and industries, for both the engineer and the student.

DESIGN AND DEVELOPMENT OF EFFICIENT ENERGY SYSTEMS, Edited by Suman Lata Tripathi, Dushyant Kumar Singh, Sanjeevikumar Padmanaban, and P. Raja, ISBN 9781119761631. Covering the concepts and fundamentals of efficient energy systems, this volume, written and edited by a global team of experts, also goes into the practical applications that can be utilized across multiple industries, for both the engineer and the student.

Electrical and Electronic Devices, Circuits, and Materials: Technical Challenges and Solutions, Edited by Suman Lata Tripathi, Parvej Ahmad Alvi, and Umashankar Subramaniam, ISBN: 9781119750369. Covering every aspect of the design and improvement needed for solid-state electronic devices and circuit and their reliability issues, this new volume also includes overall system design for all kinds of analog and digital applications and developments in power systems.

Green Energy: Solar Energy, Photovoltaics, and Smart Cities, Edited by Suman Lata Tripathi and Sanjeevikumar Padmanaban, ISBN: 9781119760764. Covering the concepts and fundamentals of green energy, this volume, written and edited by a global team of experts, also goes into the practical applications that can be utilized across multiple industries, for both the engineer and the student.

Check out these other related titles from Scrivener Publishing

HYDROGEN ELECTRICAL VEHICLES, Edited by Mehmet Sankir and Nurdan Demirci Sankir, ISBN: 9781394166381. Hydrogen electrical vehicles are an essential component of the "Green New Deal" and this book covers cutting-edge technologies designed for fuel-cell-powered cars.

PAVEMENT ASSET MANAGEMENT, By Ralph Haas, W.R. Hudson, and Lynne Cowe Falls, ISBN: 9781119038702. Whether you are new to managing pavement assets or a seasoned practitioner, this book provides the most up-to-date information in the field on the history, technology, practice and look ahead.

Printed in the USA/Agawam, MA
September 16, 2026

910661.057